Space and Society

The Space and Society series explores a broad range of topics in astronomy and the space sciences from the perspectives of the social sciences, humanities, and the arts. As humankind gains an increasingly sophisticated understanding of the structure and evolution of the universe, critical issues arise about the societal implications of this new knowledge. Similarly, as we conduct ever more ambitious missions into space, questions arise about the meaning and significance of our exploration of the solar system and beyond. These and related issues are addressed in books published in this series. Our authors and contributors include scholars from disciplines including but not limited to anthropology, architecture, art, environmental studies, ethics, history, law, literature, philosophy, psychology, religious studies, and sociology. To foster a constructive dialogue between these researchers and the scientists and engineers who seek to understand and explore humankind's cosmic context, the Space and Society series publishes work that is relevant to those engaged in astronomy and the space sciences, while also being of interest to scholars from the author's primary discipline. For example, a book on the anthropology of space exploration in this series benefits individuals and organizations responsible for space missions, while also providing insights of interest to anthropologists. The monographs and edited volumes in the series are academic works that target interdisciplinary professional or scholarly audiences. Space enthusiasts with basic background knowledge will also find works accessible to them.

The series was founded by Springer together with Founding Editor **Douglas A. Vakoch**.

Anthi Koskina • Manolis Plionis

Greek Space Law

Collaborating with a Growing Space-faring Country

Springer

Anthi Koskina
National Observatory of Athens
Athens, Greece

Department of International
Studies - School of Law
National and Kapodistrian
University of Athens
Athens, Greece

Collège IdEF - Université Sorbonne
Paris Nord
Athens, Greece

Manolis Plionis
National Observatory of Athens
Athens, Greece

Department of Physics
Aristotle University of Thessaloniki
Thessaloniki, Greece

CERIDES—Excellence in Innovation and
Technology
European University of Cyprus
Nicosia, Cyprus

ISSN 2199-3882 ISSN 2199-3890 (electronic)
Space and Society
ISBN 978-3-031-72812-9 ISBN 978-3-031-72813-6 (eBook)
https://doi.org/10.1007/978-3-031-72813-6

This Springer imprint is published by the registered company Springer Nature Switzerland AG
The registered company address is: Gewerbestrasse 11, 6330 Cham, Switzerland

The adoption of an organized system of rules, coupled with the creation of relevant official institutions, signifies a clear recognition by the State of a community's initiative and efforts to promote a valuable social good.

This book is dedicated to Greek scientists, in Greece and abroad, and especially to those involved in outer space exploration and space research in general–along with their families, friends and colleagues around the world–for their immense contribution to this field.

Foreword

The field of space law has undergone significant transformation since the launch of the first satellites, Sputnik 1 and Explorer 1, in the late 1950s. Greece, as a nation with a growing stake in space activities, has been actively involved over the years in space developments, shaping its legal framework to align with national needs and aspirations as well as international and European standards.

This book, "Greek Space Law," offers a comprehensive examination of Greece's journey in space law, from its initial steps in ratifying major international space treaties to its active participation in regional and global space initiatives. Greece's commitment to engage in the European and EU Space initiatives and providing reliable, safe, and secure space-related services is reflected in its robust legal framework, which is continuously evolving to meet new challenges and opportunities.

The volume at hand provides an in-depth analysis of the legal and policy landscape, detailing how Greece navigates its obligations under international and EU law while fostering national capabilities and infrastructure. It explores key milestones, such as Greece's accession to major international space organizations, its integration into European and EU space initiatives, and the establishment of a national framework for space activities. Through detailed chapters, the book highlights Greece's strategic initiatives, legislative measures, and the collaborative efforts that have positioned it as a significant player in the space sector.

It aims to serve as a valuable resource for policymakers, legal professionals, and scholars interested in understanding the intricacies of Greek space legislative framework and its impact on broader Greek space governance. It also underscores the importance of a well-defined legal framework in promoting sustainable and responsible space activities, ensuring that Greece remains an active state in space endeavors.

Looking toward the future, the continued development and adaptation of space law will be crucial in addressing the dynamic and complex challenges of space activities, in particular when the EU Space law is published. This book not only documents Greece's regulatory journey but also sets the stage for future advancements and collaborations in the ever-evolving domain of space.

Christina Giannopapa

Vice Chair of the Space Security
Committee of the International
Astronautical Federation (IAF), Paris,
France

Eindhoven University of Technology,
Eindhoven, The Netherlands

Preface

The present study, entitled "Greek space law: Collaborating with a growing space-faring country," was designed and conducted within the framework of academic research and funded by the program of the National Observatory of Athens: National Recovery & Resilience Plan—GREECE 2.0, id 16624, SUB7. All the opinions expressed therein belong solely to the authors and do not reflect the views of any academic or research institution, or any State or non-State official or non-official body, that is dedicated or in any manner involved in any kind of space activity. Furthermore, the information provided in this book is the result of academic research (until the time of preparation of this book) and analysis; the authors present the legislation in force and thus have no responsibility for the reality on the ground.

The Greek laws and measures on space activity are all examined and presented following a chronological order, as priority was given to the analysis of key provisions taking into account the time of their enactment. As a result, the authors wish to clarify that the present study aims at opening the debate and proposing a framework for discussing this topic and does not pretend to be an exhaustive work. It is therefore open to additions and clarifications, as well as to scientific debate on the different aspects of the Greek space legislation and its multiple dimensions.

All the documents and information produced in this study are clearly cited, to allow any person interested to have access to the sources used. More precisely, the analysis of the Greek space legislation was conducted through the thorough examination of publicly available primary sources of law, such as legislation, Presidential Decrees, ministerial decisions and declarations, whereas related preparatory documents (e.g., Explanatory Memoranda, Verbatim Reports of Proceedings, etc.) were used to clarify and explain the rationale behind important provisions applying to space activity. Consideration was also given to the currently available scholarly sources and Greek bibliography on some of the major matters dealt with in this study. Finally, press articles on the Greek space policy, space programs, as well as projects and technological or research developments are being presented, as complementary means for the interpretation of the provisions that have been adopted in the domestic legal order.

To provide a general but accurate view of the Greek space legislation,[1] the research is divided into two parts. Following a brief introduction, the first part presents the Greek legislation allowing the country to enter into international agreements and organizations involved in space activity, while the second part focuses on the content of the national space law, namely law n° 4508/2017, as part of the national framework applying to space activity that is carried out under the jurisdiction of the Greek State. There, the competent authorities and bodies are examined first; subsequently, the topics and issues addressed by the Greek space law are referred to and analyzed, while elements of comparative law are presented (where necessary) for the assessment of the Greek space law provisions.

Eventually, the study contains a verbatim translation of the Greek space law (law n° 4508/2017) in English, in Annex I. The verbatim translation was chosen so as to avoid any form of misleading expressions in the English version of the document, potentially leading to a misinterpretation or misunderstanding of the Greek space law provisions. Said translation has only been adapted to a limited extent, and mainly with the aim to be sufficient for the reader to understand the basic intent (i.e., the *rationale*), the subject matter, and the content of each article of the Greek legislation.

Athens, Greece

Anthi Koskina
Manolis Plionis

[1] The aim of the study is not to replicate or analyze international law discussions on space activity matters; the in-depth evaluative analysis of international space law obligations of States is, in truth, beyond the scope of this project. Therefore, only a few books on international space law are being mentioned and they have been used solely as a point of reference.

Contents

Chapter 1
Introduction

According to the Hellenic Republic Statement submitted to the 58th session of the Scientific and Technical Subcommittee of the UN COPUOS in 2021 [1], Greece became a space-faring country in 2003, when the first Greek Telecommunication Satellite was launched under the name *Hellas Sat 2*. *Stricto sensu*, said statement is accurate [2];[1] however, it fails to adequately reflect the emergence and the development of a Greek space activity *lato sensu*, or the progresses made in the field of space policy long before the launching of *Hellas Sat 2*.

In reality, Greece got involved in space activity and policy much earlier than that, by ensuring first that the principal international space law treaties and agreements were incorporated into national legislation, namely following the adoption of the necessary *legislative and regulatory measures* (from the 70s onwards); and secondly by *entering into membership* and associated membership of several international (and regional) organizations in this field, and participating in space-related research and development programmes to encourage and promote the domestic space activity. Following on from that, it successfully developed its own space sector and space research projects, while introducing new legislation to create an adequate legal framework within which space operations would be implemented.

By way of background, it is appropriate to briefly recall that the international space law regime was laid down following the successful launching of the Soviet satellite *Sputnik 1* in 1957, and the American one in 1958 (*Explorer 1*). However, after the poor ratification record of the Moon Agreement adopted in 1979 [3], the legislative activity in the field of international space law declined, leaving room for space actors to elaborate and adopt a wide array of soft-law instruments. Hence, the emergence of new space-faring countries played a crucial role in defining priorities in space activity, and paved the way for an increase in regional and bottom-up initiatives.

---

[1] A space faring nation is one that is simply defined as a State "having vehicles capable of traveling beyond the earth's atmosphere" [2].

A. Koskina, M. Plionis, *Greek Space Law*, Space and Society,
https://doi.org/10.1007/978-3-031-72813-6_1

In this context, the Greek participation in space law and space policy developments could be divided into three phases: a first phase during which international space law is incorporated into the Greek legal order; a second one in which the Greek State mainly acts in its capacity as an EU Member State, embracing policy developments at EU level and joining EU space initiatives; and a third one where the emphasis is finally put on establishing a national framework for space activity and on promoting domestic space operators.

1.1 Shifting from the International to the Regional, and then to the National Level

The international space law regime was debated and adopted *first* on a global level, in the context of international space law institutions and bodies. Indeed, policy agendas and normative frameworks were initially decided within international organizations active in this field, where only a few nations involved then in space activity were setting the tone. More precisely, space activities started within the framework of the Cold War (1947–1991)—under the lead of the US and the USSR [4] which were politically and ideologically fundamentally opposite to each other [5]—and were seen as a sign of political strength. Taking this context into account, the two principal space-faring nations of that time committed to the peaceful exploration and use of outer space, in line with the principles of non-proliferation of nuclear weapons and the sustainable use of outer space resources (as a means to avoid any outbreak of hostilities in space), and they created the global institutions that were necessary to advance this specific policy (i.e., international bodies for space law). Hence, a range of space-related organizations were set up, as expected, in the decades following the launching of the first satellites—*Sputnik 1,* in 1957 and *Explorer 1*, in 1958—namely from the late 50s to the late 70s. With a time lag of a few years, many countries of that period that were not yet actively involved in space (such as Greece) participated in these important developments essentially by acceding to these international space law instruments and by joining said institutions. This phase spread for Greece over, roughly, the 70s and the 80s.

Progressively, new space-faring countries and regional actors emerged; as a result, decision- and policymaking was now increasingly being developed at regional level—i.e., at the EU level for the Greek State—, to accompany and encourage the emergence of ambitious regional space stakeholders, as well as new industry sectors with a strong potential for growth. Greece moved then to the next phase, leaving behind its sporadic involvement with space and space-related activities (which had been characterized by the enactment of laws that were mostly aimed at ensuring acceptance of international space law and all the space activity related standards). Indeed, the late 80s and the 90s, are marked—for Greece—by the emergence of its own space sector and an increased involvement in regional space-related activities, as it then joined and seeked to actively participate in more and more EU initiatives. Along the years, and mostly since the 2000s, the EU developed a framework aimed at meeting the challenges of a new space era, by increasing its autonomy and competitiveness in the space sector, and at the same time gave its member States the opportunity to take more action in this field. As a result, participating in a growing number of EU space initiatives allowed Greece to acquire additional knowledge on space matters and to develop a domestic space industry; in the late 2010s, it finally adopted its national space legislation and created a national space authority, paving the way for even more efficient cooperation and involvement in this area.

1.2 Legal Basis for Incorporating International and EU Rules into Greek Law

The Greek State has fully aligned its legislation with the international (space) law obligations that it has undertaken, as provided in its own constitution. Indeed, the Greek constitution—i.e., the supreme law of the Greek State—lays down that all "*The generally recognized rules of international law, as well as international conventions as of the time they are ratified by statute and become operative according to their respective conditions, shall be an integral part of domestic Greek law and shall prevail over any contrary provision of the law. The rules of international law and of international conventions shall be applicable to aliens only under the condition of reciprocity*" (Art. 28 para. 1) [6].

In addition to that, a more specific constitutional provision was adopted, to facilitate the transposition of EU legislation. Hence, to clarify the scope of Article 28 mentioned above, its *Interpretative Clause* reads "*Article 28 constitutes the foundation for the participation of the Country in the European integration process*".

Therefore, any international space law treaty and/or agreement which has been duly signed and ratified by the Greek government becomes part of the national legal system and a binding higher source of law, in line with Art. 28 par. 1 of the Greek Constitution. Following on from that, all the international space law instruments which had been adopted by the international community with the aim to create a regulatory framework that would meet the desires and demands of the multiplying stakeholders involved—and would as well promote dialogue and cooperation in all areas of space activity—, were taken into account to develop the Greek legal framework.

In many cases, however, international (space) law treaties are binding upon States parties as to the result that must be achieved, but still require the adoption of additional implementing measures; from a different angle, although necessary, the adoption of national (space) law is neither self-evident, nor nature given [7]. Hence, considering the practical need to encourage the adoption of harmonized national legislations in line with international law, in addition to creating correspondent administrative authorities and bodies, non-binding—i.e., soft law—instruments were adopted to be used as a basis. In particular, the *Resolution on Recommendations on National Legislation Relevant to the Peaceful Exploration and Use of Outer Space* released in 2013 [8], and the 2012 *Sofia Guidelines for a Model Law on National Space Legislation of the International Law Association* (or *Sofia Guidelines*) [9], provided significant help to States in meeting their commitments under international space law.

Following on from the above, it appears that the Greek State's regulatory efforts were essentially framed by the tools and objectives adopted within the context of the international space law treaties and the regional (EU) space laws and policies, which Greece made it best efforts to transpose into national law. Be that as it may, the times are changing and experts already make reference to a *new space era*, characterized by revolutionary developments in the field of space robotics—also based on the growing use of Artificial Intelligence—, and the development and increased use of different kinds of small spacecrafts (ex. CubeSats; mega-constellations) that are reliable, economically viable and affordable by all, including non-established space-faring countries.

Consequently, Greece took this opportunity and further increased its involvement in space activity, while shifting its focus on internal developments to provide attractive framework conditions for space operators, based on transparency and legal certainty.

The present work aims to give an overview of all these developments. Therefore, it is divided into two main parts, namely Part I on the accession to international space law instruments, including the joining of international initiatives and the transposition of EU laws. As a result of that, Part II puts the emphasis precisely on measures adopted at the national level that were put in place for implementing the Greek State's international obligations, which in truth went hand in hand with successes achieved in the field of space research and the emergence of a Greek space sector.

References

1. UNCOPUOS STSC. (2021). *Statements to the 945th meeting of the scientific and technical subcommittee, Monday, 26 April 2021 - Item 3: General exchange of views and introduction of reports submitted on national activities (Greece).* Accessed from https://www.unoosa.org/oosa/en/ourwork/copuos/stsc/2021/statements.html
2. Merriam-Webster. (n.d.). Spacefaring. In *Merriam-Webster.com dictionary.* Accessed July 15, 2024, from https://www.merriam-webster.com/dictionary/spacefaring
3. United Nations Office for Outer Space Affairs. (1979). *Agreement governing the activities of states on the moon and other celestial bodies* (the Moon Agreement) (adopted 18 December 1979, entered into force 11 July 1984). *United Nations Treaty Series, 1363,* 3. Retrieved from https://www.unoosa.org/oosa/en/ourwork/spacelaw/treaties/moonagreement.html
4. Ehrenfreund, P., & Peter, N. (2009). Toward a paradigm shift in managing future global space exploration endeavors. *Space Policy, 25*(4), 244–256.

5. Rao, U. R., & Chandrasekhar, S. (1983). International issues in remote sensing. *Proceedings of the Indian Academy of Sciences Section C: Engineering Sciences, 6*, 373–386. Accessed from https://link.springer.com/article/10.1007/BF02881140
6. The Constitution of Greece. *Revised by resolution of November 25, 2019 of the IXth Revisionary Parliament.* Accessed from https://www.hellenicparliament.gr/UserFiles/f3c70a23-7696-49db-9148-f24dce6a27c8/THE%20CONSTITUTION%20OF%20GREECE.pdf
7. von der Dunk, F. (2006). Fundamental provisions for National Space Laws. In *Meeting international responsibilities and addressing domestic needs: Proceedings United Nations, Vienna, 2006.* Accessed from https://digitalcommons.unl.edu/cgi/viewcontent.cgi?article=1010&context=spacelaw
8. UNGA. (2013). *Resolution 68/74 adopted by the general assembly on 11 December 2013. Recommendations on National Legislation relevant to the peaceful exploration and use of outer space.* UN Doc A/RES/68/74. Accessed from https://www.unoosa.org/pdf/gares/A_RES_68_074E.pdf
9. UNCOPUOS LSC. (2013). *52nd session. Information on the activities of international intergovernmental and non-governmental organizations relating to space law.* UN Doc A /AC.105/C.2/2013/CRP.6. Accessed from https://www.unoosa.org/pdf/limited/c2/AC105_C2_2013_CRP06E.pdf

Part I
Accessing International and EU Initiatives

The legislation applying to space activities being conducted under the responsibility of the Greek State was developed to take into account international (and regional) space laws, purposes and priorities. In other words, as the first rules on space activity were the ones adopted at the international level—by a small number of States participating in space activity at that time—, the first characteristic of the Greek legal framework is that its initial steps focused on aligning with international efforts, at all levels; which eventually paved the way for the development of a national framework that would best contribute to the promotion of national research efforts and encourage the development of a Greek space sector.

Against this background, Part I aims at presenting the principal steps that were taken in order to ensure the accession to international space law instruments and space-related organizations, including international space law *fora* (Chap. 2). Thus, more precisely, details are provided in relation to the ratification of international space law treaties which started from the 70s (Sect. 2.1); accession to the major international law organizations (Sect. 2.2) and—finally—the consideration of several soft-law instruments that are increasingly being adopted to regulate more technical aspects of space activity (Sect. 2.3).

Following on from that, a whole chapter is dedicated to presenting the impetus that was gained from the Greek State's participation in numerous EU space related activities together with its involvement in regional (EU) space programmes (Chap. 3). In particular, details are provided regarding its participation in EU organizational structures for space services, and steps taken to join the ESA (Sect. 3.1); as a result, all this essentially allowed Greece to develop its own expertise and capacities in the fields of space activity (Sect. 3.2).

Chapter 2
Accession to International Space Law Instruments and Organizations

The first membership of Greece to an international space-related organization is said to be its admission to the International Telecommunication Union (ITU). Indeed, pursuant to the ITU official homepage, the organization "was founded in Paris in 1865 as the International Telegraph Union. It took its present name in 1932, and in 1947 became a specialized agency of the United Nations"; thus its founding act—namely, the *International Telegraph Convention* dated 17 May 1865 [1]—was signed by 20 States, including Greece (the Greek government was then headed by the king[1]) which makes it a founding member (see also, the statement of the Hellenic Republic at the UNCOPUOS in 2021 [2]).

However, the establishment and accession to the ITU on 1 January 1866 [3] cannot be regarded as a participation in a space-related organization *stricto sensu*, given that the ITU initially aimed at "standardizing *telegraphy equipment*, setting uniform operating instructions, and laying down common international tariff and accounting rules" [4] (emphasis added); in other words, it was already well into 1959 when the ITU focused on space matters, by setting up a study group responsible for studying space radiocommunications, namely Study Group IV of Space systems and Radio astronomy [5].

Hence, in reality, the *first* participation of Greece in the space-related international *fora*, organizations and bodies that were initially set up—i.e., starting from the late 50s—essentially took place *around the 80s* (Sect. 2.2). Interestingly though, a decade before this wave of accessions, namely *in the 70s*, the Greek government had already ratified the principal international space law treaties and agreements negotiated and signed within the context of said international organizations (Sect. 2.1); critically, it evidently kept, and still keeps taking into account all the developments in international space law including as regards soft law (Sect. 2.3).

[1] The International Telegraph Convention makes reference to *His Majesty the King of the Hellenes;* see [1], (emphasis added).

A. Koskina, M. Plionis, *Greek Space Law*, Space and Society,
https://doi.org/10.1007/978-3-031-72813-6_2

2.1 Ratification—in the 70s—of the Major International Space Law Treaties

As a result of the successful completion of the first space activities back in the late 50s, a new body of international law emerged to set out the basic principles that would govern human endeavors in outer space. Be that as it may, it must be noted, for the sake of precision, that the initial response from the international community was (at that time) to first of all create international bodies allowing for constructive dialogue, and able to provide an efficient framework for the negotiation of rules that would regulate the conduct of space activities.

In other words, it was the UN COPUOS back in 1958—along with the UN OOSA, in 1959—that was originally created, as a structure within which the fundamental rules governing activities in outer space would be negotiated and agreed upon by States. Nonetheless, given that the Greek government first ratified the international space law instruments, and acceded to international space law organizations only in the aftermath of said ratifications, the first topic (namely, the ratification of international space law treaties) will be presented first, to follow the chronological order of these developments.

2.1.1 Key Elements of the Principal International Space Law Treaties

Initial steps towards the creation of a specific regulatory framework at international level included adoption of the *Declaration of Legal Principles Governing the Activities of States in the Exploration and Use of Outer Space*, in 1963 [6] (in this instrument, the guiding principles for all activities pertaining to the exploration and use of outer space were laid down, and they subsequently formed the basis of the Outer Space Treaty adopted a few years later), as well as the adoption of the *Nuclear Test Ban Treaty*, also in 1963, which aimed at achieving "the discontinuance of all test explosions of nuclear weapons for all time, continue negotiations to this end, and put an end to the contamination of man's environment by radioactive substances" [7: Preamble].

Following on from that, the *Treaty on Principles Governing the Activities of States in the Exploration and Use of Outer Space, including the Moon and Other Celestial Bodies*, referred to as the *Outer Space Treaty*—hereafter OST—was enacted by the UN General Assembly, in its Resolution 2222 (XXI) [8]. Given the wide ratification of the OST, said treaty is regarded to be the cornerstone of international space law and legally binding on literally all States, due to both its high level of recognition and its customary character [9].

In brief, the OST laid down the principle of the freedom of exploration and use of outer space to the benefit of all countries, irrespective of their degree of economic or scientific development (OST, Art. I); the principle of non-appropriation (OST, Art.

II) applying to space and space resources, including the Moon and other Celestial bodies; and the obligation to conduct space activities in accordance with International Law, including the Charter of the United Nations. In addition to that, all member States agreed to act with regard to the purpose of maintaining international peace and security—see OST, Art. III: "States Parties to the Treaty shall carry on activities in the exploration and use of outer space (. . .) in the interest of maintaining international peace and security and promoting international cooperation and understanding"; and Art. IV para 1: "States Parties to the Treaty undertake not to place in orbit around the Earth any objects carrying nuclear weapons or any other kinds of weapons of mass destruction (. . .)" and para. 2 "The Moon and other celestial bodies shall be used by all States Parties to the Treaty exclusively for peaceful purposes (. . .)"—, and to bear international responsibility for their activities conducted in outer space (OST, Art. VI: "States Parties to the Treaty shall bear international responsibility for national activities in outer space, including the Moon and other celestial bodies, whether such activities are carried on by governmental agencies or by non-governmental entities, and for assuring that national activities are carried out in conformity with the provisions set forth in the present Treaty (. . .)").

The OST was complemented by specific treaties, that regulate particular aspects of outer space activities. Hence, in April 1968, the *Agreement on the Rescue of Astronauts, the Return of Astronauts and the Return of Objects Launched into Outer Space* (*Rescue Agreement*) was adopted [10], to ensure the safety of humans in space by "rendering all possible assistance to astronauts in the event of accident, distress or emergency landing, the prompt and safe return of astronauts, and the return of objects launched into outer space" [10, see Preamble]. In March 1972, the *Convention on International Liability for Damage Caused by Space Objects* (*Liability Convention*) established international rules and procedures for the recovery of damages for losses caused by space objects [11]. In January 1975, the *Convention on Registration of Objects Launched into Outer Space* (*Registration Convention*) created a system for the mandatory registration of all objects launched in outer space to assist States in their identification [12] and finally, the *Agreement Governing the Activities of States on the Moon and Other Celestial Bodies*, signed in December 1979 and entered into force in 1984 (*Moon Agreement*)—based on key rules that had been established in the previous international law instruments, and mainly in the OST—, aimed at regulating the exploration and use of the Moon and other celestial bodies [13].

The Greek State ratified four of the five international space law treaties mentioned above. Hence, it is bound by all the relevant rules of international law pursuant to Art. 28 para. 1 of the Greek Constitution, and ensures compliance with said framework.

2.1.2 Greek Laws Pertaining to the Ratification of International Space Law Treaties

As previously mentioned, the ratification of the principal space law treaties took mainly place in the 70s. More precisely, the Greek State adopted then the following laws and legislative decrees, ratifying said treaties and agreements, all of them duly published in the Greek Government Gazette:

(1) The *Treaty on Principles Governing the Activities of States in the Exploration and Use of Outer Space, including the Moon and Other Celestial Bodies* (hereafter, OST), adopted on 27 January 1967, was signed by the Greek government on the same date and ratified on 19 January 1971 [14]. The OST was received into Greek law by means of the Legislative Decree n° 670/1970, published in the Greek Government Gazette, Volume A '208 [15]. As a result, the Legislative Decree n° 670/1970 was the first instrument of reference, creating legal terminology for activities aimed at the exploration and use of outer space—i.e., space activities and related operations—, *via* the official translation of the OST provisions in Greek. Key concepts, related to *inter alia* the nature of the right of States to explore and use outer space, were for the first time translated and introduced into domestic law.

Be that as it may, translating a legal document into a different language is inherently risky. In reality, this is all the more true when terms—such as those related to space activity— needed to be understood (and translated) in a legal context which was not yet familiar with outer space operations, as was the case for Greece in the 70s. As a result, OST, Art. I, mentioning that "*The exploration and use of Outer Space (...) shall be the province of all mankind*" (emphasis added) was translated into "The exploration and use of space (...) *comes under the jurisdiction of all mankind*" (emphasis added), which is a more specific—and better defined—legal concept: i.e., in Greek language, the verbatim translation, mentioned in Legislative Decree n° 670/1970 on the Ratification of the Outer Space Treaty, is "(...) *θέλει δε ανήκει εις την δικαιοδοσίαν (PROVINCE) ολοκλήρου της ανθρωπότητος*" [15: Art. I translated]. From this perspective, it may be argued that the meaning assigned in the Greek translation of the OST to the wording of the rule adopted in Art. I of the OST could be regarded as being slightly different and eventually narrower.[2]

[2]The term "province of all mankind" is to a certain extent vaguer and more general than the concept of "(being) under the jurisdiction (of an entity)". *Inter alia*, "be somebody's province" is explained in the Cambridge Advanced Learner's Dictionary and Thesaurus [16] as "a subject or activity of special interest, knowledge, or responsibility:" Similarly, being the "province of" is defined by the Macmillan Dictionary, as "a subject that someone knows about or is responsible for" (ex: The family finances are his wife's province) [17]. On the contrary, the concept of being "under the jurisdiction of" is clearer and well defined in international law, arguably focusing on the implementation of the rule of law (in this case, of International Space Law) rather than on general policy-making; see, for instance "(D)efinitions stating in more or less varying terms, but substantially, that jurisdiction is the power to hear and determine a cause; the other class, which is small, stating substantially that jurisdiction is the right to hear and determine a cause" [18].

In theory, it may occur that the terminology or meaning adopted in national laws appear to run against the legal meaning of the authentic version of an international treaty; still, such laws must in any case be interpreted in the light of Art. 31 (i.e., "A treaty shall be interpreted in good faith in accordance with the ordinary meaning to be given to the terms of the treaty in their context and in the light of its object and purpose (...)") and Art. 32 ("Recourse may be had to supplementary means of interpretation, including the preparatory work of the treaty and the circumstances of its conclusion, in order to confirm the meaning (...)") of the *Vienna Convention on the Law of Treaties*, signed in 1969 [19]. Hence, in case of uncertainties arising due to divergence between the different language versions, the authentic versions of the OST—which are precisely referred to in the OST, Article XVII: "This Treaty, of which the English, Russian, French, Spanish and Chinese texts are equally authentic, shall be deposited in the archives of the Depositary Governments"—remain the only legally binding ones, while all versions presumably have the same meaning. Indeed, Art. 33.3 of the Vienna Convention "states that the treaty terms are presumed to have the same meaning in each authentic text ("the presumption"), thus making it unnecessary to compare language versions on a routine basis" [20]. In other words, regardless of the concepts adopted in national versions (e.g., in the ratification acts), the meaning of any translated term cannot be different from the authentic versions' one. Therefore, States parties (including Greece, as a party to the treaty) are bound to apply the provisions laid down in treaties, such as the *Outer Space Treaty* (OST) in this spirit as well as in good faith, as laid down in Art. 31.1 of the Vienna Convention [19].

From a different perspective, legal experts are also in favor of considering that many OST articles have now entered into force as customary rules of international law and are therefore binding on all members of the international community, independently of whether these have signed and ratified the treaty. This approach is based on the wide acceptance of the OST—now ratified by more than 110 States [14]—, and on the fact that many of the rules established therein were the most common denominator of practices already existing among the space faring countries at that time. Following this reasoning, namely from the viewpoint that the OST mainly reflected existing practices, considering the content of said practices in the field of space activity may further our understanding of what "implementing the OST in good faith" means, and eventually provides a solution in the case where divergences in legal interpretations would occur (to our knowledge, no problem regarding the implementation of the OST by the Greek State has occurred up to now).

(2) The *Agreement on the Rescue of Astronauts, the Return of Astronauts and the Return of Objects Launched into Outer Space* (the "*Rescue Agreement*"), agreed upon on 22 April 1968 [10], was signed by the Greek government on 11 June 1968, and ratified on 7 July 1975 [21]. The Rescue Agreement was received into Greek law by means of Legislative Decree n° 189/1974, published in the Greek Government Gazette A '350 (pp. 2572a et seq.) [22]. This second Legislative Decree n° 189/1974 kept the structure of Legislative Decree n° 670/1970 [15] transposing the OST into

the Greek legal order. In other words, an official *verbatim* translation of the Rescue Agreement is provided together with the English version of the Treaty.

(3) In the wake of these developments, the *Convention on International Liability for Damage Caused by Space Objects* (the "*Liability Convention*") adopted on 29 March 1972 [11], was signed by the Greek Government on 27 April 1977, and ratified on 4 May 1977 [23]. The Liability Convention was transposed into domestic law through Law n° 563/1977, published in the Greek Government Gazette A '75 (pp. 595 et seq.) [24], and likewise kept a structure which was similar to that of the former Legislative Decrees transposing first the OST and then the Rescue Agreement (namely, an official *verbatim* translation of the Liability Convention is provided together with its English version).

(4) Interestingly, contrary to the above, the *Convention on Registration of Objects Launched into Outer Space* (the "*Registration Convention*"), adopted on 14 January 1975 [12] was only signed by the Greek State in the 2000s. More precisely, it signed the Registration Convention on 27 May 2003 [25]; the convention was then ratified on 4 April 2003, through Law n° 3131/2003, published in the Greek Government Gazette A '79 (pp. 1279 et seq.) [26] and followed, once again, the same structure as the acts ratifying the previous international space law treaties.

Greece has not ratified—up until this date—the last of the five major international space law treaties, which is the *Moon Agreement* adopted in 1979 that was opened for signature on 18 December 1979 [27]. However, in reality, only very few states (i.e., around 18, until now) signed or ratified the Moon Agreement.

One explanation of this could be that, although for a long period of time, governments had played the principal role in space economy—acting as investors, and as operators and customers likewise—, space activity steadily gained in importance and attracted more and more private actors showing growing interest and participation. Parallel to that, the State-centric approach to space law was increasingly called into question. As a result, only a few States ratified the Moon Agreement, while no additional binding instrument of international space law was ever concluded, either through the UN COPUOS or other similar *fora*, since said treaty was adopted in 1979. Put differently, in a world where competition for space matters was growing, soft law rules and codes of conduct (that are not binding but do, nevertheless, produce indirect legal results) became a principal feature of the current normative framework applying to space activity; hence, it was underlined that "There has been a strong tendency towards the development of soft law guidelines and 'codes of conduct' for space-related matters, notwithstanding the inherent risks that this (potentially) brings of greater 'non-compliance" [28].

Soft-law instruments are, evidently, taken into account as appropriate by the Greek State, when deciding upon space matters. However, such measures are a characteristic of the last few decades. Therefore, the steps taken as regards the accession of the Greek State to the international space law bodies and institutions in the 80s will be presented first, to make the progresses made at national level more readable.

2.2 Accession to the Main International Space Law Organizations During the 80s

Greece has, traditionally, had a strong and dynamic presence on the international scene; indeed, it was, by way of illustration, an inaugural member of the United Nations [29], as it signed—on 26 June 1945—the founding Charter of the UN [30]. As regards international space law in particular, the Greek State ratified first the key international space law treaties (which mainly took place in the 70s), and soon took a major step forward by accessing the international and intergovernmental organizations that were already active in the field of space law and policy, as well.

More precisely, Greece acceded first (in 1980) to the principal *forum* that had been established for dialogue and international cooperation—in reference to space matters—at the policy level, the UN Committee on the Peaceful Uses of Outer Space (UN COPUOS). Following that, it additionally acceded to several international bodies which had been created to provide for more practical arrangements as regards the use of outer space and space resources, such as mainly to manage constellations of satellites (or any other types of space infrastructure being used at that time) providing international communication and broadcast services; this second category of international bodies essentially comprised entities such as INTELSAT and INMARSAT.

2.2.1 *Policy Making: Joining the UN COPUOS in 1980*

The Ad Hoc Committee on the Peaceful Uses of Outer Space (UN COPUOS) was created in 1958 [31], and converted into a permanent body in 1959 [32], to "review (. . .) the area of international cooperation, and (. . .) study practical and feasible means for giving effect to programs in the peaceful uses of outer space which could appropriately be undertaken under United Nations auspices", and to "study the nature of legal problems which may arise from the exploration of outer space" [*Ibid.*]. In the same year—i.e., in 1958—the United Nations Office for Outer Space Affairs (UN OOSA) was created as an office (that is, a small expert unit) within the UN Secretariat, to service, assist and advise the UN COPUOS [33].

Greece was appointed as a member of the UN COPUOS in 1980, as mentioned in the UN GA Resolution 35/16 [34]; since then, it is regularly participating in the works of its committees, through timely, adequate and targeted interventions which allow the Greek State to also strengthen its international position. To make its involvement more readable, the following paragraphs give an overview of the Greek State's participation as well as of its main domains of intervention at the UN COPUOS Legal Subcommittee, presenting selected initiatives which were put forward in recent years (i.e., rather than a detailed report of the work carried out by the Greek delegation in said organization, so as not to tire the reader).

2.2.1.1 Initiatives Aimed at Rethinking Specific Rules of International Space Law

(1) Greece has co-sponsored a proposal that was made by the Czech Republic as to the *reviewing of existing norms of international law applicable to space debris*. Hence, given that during the past decades, the creation of man-made space debris became one of the most discussed issues relating to the further development of outer space activities, "(t)he Czech Republic was one of the member States of the Committee that drew the attention of the Legal Subcommittee to those problems in a background note on the review of existing norms of international law applicable to space debris as an initial step in the consideration of the legal aspects of space debris". Due to the importance of this topic, the proposal was actively supported by Greece [35].

(2) The Greek State submitted, together with China and the Russian Federation, a proposal on the desirability of drafting a *universal comprehensive convention on international space law* [36]. In this case, the suggestion was to reshape international space law, so as to take into account the increase in the number of space-faring countries, as well as the growing participation of private stakeholders. According to this proposal, "certain provisions of the existing basic space treaties should be adapted or modified to take into account current realities"; thus, building on previous efforts that had been made, the suggestion was to review "the issue of the appropriateness and desirability of drafting a universal comprehensive convention on international space law as a sub-item of the agenda item entitled "Status and application of the five United Nations treaties on outer space",

(3) Together with Brazil, Greece proposed to include a new item in the agenda of the UN COPUOS Legal Subcommittee, under the title "Discussion on the development of *an international convention on remote sensing*" (emphasis added) [37]. Practically, the suggestion was to review the *Principles Governing the Use by States of Artificial Earth Satellites for International Direct Television Broadcasting*, with a view to possibly transforming the text into a treaty in the future that would take new circumstances into account [*Ibid.*],

(4) Greece submitted (jointly with several other States) a proposal to the UN COPUOS aimed at considering a new agenda item, that would be entitled "Practice of States and international organizations in registering space objects" [38]. Following on from this proposal, the *Working Group on the Practice of States and International Organizations in Registering Space Objects* was created, to put focus on specific aspects such as the harmonization of practices (both administrative and practical); the non-registration of space objects; and practice with regard to the transfer of ownership of space objects while in orbit; in addition to practices with regard to registration/non-registration of "foreign" space objects [38]. In this context, Greece also informed the Secretary-General of the United Nations—by note verbale—about the *de facto* maintenance of a national registry in 2008, namely before such registry was formally established by the Greek space law in 2017 [39],

(5) The Greek State regularly participates in the exchange of information and views, such as by communicating its position regarding the definition and delimitation of outer space. In particular, Greece notified in 2017 the UN COPUOS legal subcommittee that "(w)ith regard to the delimitation of outer space, and given the current level of space and aeronautics activities in Greece, there (is) not yet national legislation defining or delimiting outer space". Nonetheless, it additionally expressed the opinion that "(g)iven the interdependence of the issue with new technologies being developed in States, for instance in the area of suborbital flights, it is indispensable that this delimitation takes into account the existing international aeronautical regulations of ICAO. For instance, to the extent that a suborbital flight (of any kind) takes place in the airspace, the suborbital vehicle in question should, for that part of its journey, be submitted to the applicable air traffic rules (national rules or FIR rules), in order to ensure a safe, regular and efficient air transport (art. 44 (d) of the Chicago Convention)" [40].

In this context (namely, with regard to the need to define and delimit outer space),[3] the perspective was also expressed that "not only the differentiated capacity among States to exercise their sovereignty in any part of space, but also the prohibition of national appropriation by claim of sovereignty or by means of use or occupation as confirmed by Article II of the Outer Space Treaty (1967), would undeniably pose critical problems in a possible adoption of the spatial theory. Additionally, should the boundary be based in relation to anti-satellite tests or the activities of certain States to remove and destroy own satellites, it would be at least at an altitude equal to that of satellite orbits. This solution would not serve as a clear boundary between airspace and outer space". Hence, Greece expressed itself in favor of the functional approach, "given the current and foreseeable technological state[4]".

2.2.1.2 Clarifications on Matters Contingent Upon the Definition & Delimitation of Space

The Greek State reiterated the same approach on the definition and delimitation of outer space two years later, namely in 2019 [41]. On this occasion, it also expressed clear views on closely related topics, such as whether there is a relationship between plans to establish a system of space traffic management and the definition and delimitation of space. In this regard, it observed that "the issue of definition and delimitation of outer space could have an impact on plans to establish a system of

---

[3] See for more details Sect. 2.2.1.2.

[4] More analytically, "According to this theory, space is to be considered outer space at any distance from the surface of the Earth as long as it may be used by space objects, i.e., capable of performing space flight. Namely, the different nature of space activities, and the fact that there is no connection with the underlying territory, implies that these activities shall, wherever conducted, be exclusively subject to the sovereignty of the launching States. Therefore, the legal regime of outer space should be determined based on the capacity of space launches or orbits of space devices, at their lowest perigee" [40].

space traffic management. However, the international community must be prepared to provide answers in this respect, even in the absence of such a definition/ delimitation" [41].

Moreover, in reference to whether there is a relationship between suborbital flights for scientific missions in space—and/or for human transportation—and the definition and delimitation of outer space, the answer was that the two elements[5] are effectively interrelated. The Greek State adhered, again, to one of the prevailing approaches to this question;[6] indeed, it clearly highlighted that "the functionalist approach on the definition and delimitation of outer space may provide greater clarity than the spatialist one, as it would result in the application of only one legal regime (i.e., space law) to suborbital flights, whose primary function is to enter outer space (despite the fact that they are also designed to traverse airspace)[7]".

In response to whether a legal definition of suborbital flights for scientific missions and/or for human transportation will be practically useful for States (as well as for other actors) with regard to space activities, it made it clear that such step would not only be useful, but it would furthermore "contribute to the setting up of a system of global space governance" [41]; which opinion points out, at the same time, that in the context of globalization, global governance would be the right path to follow.

Be that as it may, the Greek State did not provide any definition of suborbital flights (i.e., in this document). Answering the question of how could suborbital flights be defined, it only notified that "the definition of suborbital flights for scientific missions or manned space flights has particularities, making it a complex subject that requires careful study, which we believe should be done within the competent UN bodies following a wide-ranging open consultation", and that the legislation that should apply to such activity "is related to the content of the definition of suborbital flights and to the specific characteristics of the flight". This definition, as it was rightly underlined by Greece, "will have a significant impact on the development of space law, and the extent of impact is directly related to the content of the definition of suborbital flights for scientific missions and manned space flights" [41].

[5]Namely, (a) suborbital flights for scientific missions and/or for human transportation and (b) the definition and delimitation of outer space.

[6]Which are (i) *the spatialist approach* (i.e., the definition and delimitation of outer space should take into account the technical feasibility of exercising effective control; hence, "the sovereignty of a State should extend vertically as high as it is able to enforce that sovereignty" [42: 145]) and (ii) *the functional approach* (the definition and delimitation of outer space should only take into account the intention between the activity involved [42: 150]).

[7]The Greek State explained further its position by noting that "(t)he functionalist approach distinguishes between aeronautical and astronautical activities, based on the primary purpose of a vehicle's design rather than on the location of the vehicle at any single point in time. Under this approach, a company's suborbital operations would be governed under space law only (with the exception of air traffic management rules for that portion of the flight that crosses airspace), because suborbital flights are designed for space travelling" [in 41].

2.2.1.3 Initiatives and Views in the Context of the Space Resources Exploitation debate

In recent years, major legislative developments took place (first and foremost at the national level) with respect to space resources use and exploitation; such as was the adoption of the US Commercial Space Launch Competitiveness Act in 2015 [43], and the Luxembourgish Law of 2017 on the Exploration and Use of Space Resources, laying down that outer space resources are capable of being owned [44: Art. I]. This gave rise to debates, at the international level as well [45], on whether or not such developments were compatible with the rules established in international space law (especially, Art. II of the OST). As a result, a series of discussions were soon organized at the international level within the UN COPUOS Legal Subcommittee, mainly starting from its fifty-sixth session (i.e., in 2017).

Building on previous initiatives [45: at 7],[8] Belgium and Greece proposed (during the fifty-eighth session of the Legal Subcommittee) a working paper, entitled *Proposal for the Establishment of a Working Group for the Development of an International Regime for the Utilization and Exploitation of Space Resources*. According to this proposal, such working group would have had the general mandate to develop and propose alternative legal solutions capable of providing the legal certainty necessary for acts of exploration, exploitation and utilization of outer space resources, which was supported by at least one other delegation [46].

However, the proposal prompted much discussion as well. By way of illustration, "(s)everal States, including the United States (...) and Canada (...), argued that there was no consensus amongst States regarding the desire for an international regime dedicated to space resources. Furthermore, numerous States, including Mexico (...) and Luxembourg (...), suggested that the three-year work plan was overambitious and likely unattainable" [45: 12]. In fact, although it did not prove possible to reach agreement, "there was an overarching desire to continue the discussions on the topic of space resource activities and the possibility of a dedicated Working Group" [45: 13].

As a result, a follow-up proposal was submitted in 2021, at the sixtieth session of the UN COPUOS Legal Subcommittee, for the establishment of a working group on potential legal models for activities in the exploration, exploitation and utilization of space resources [47], containing detailed work plans and methods of work. More precisely, the proposed working group is expected to submit to the Legal Subcommittee, at the end of a five-year workplan: (1) principles and practical measures for space resources activities on celestial bodies, for possible adoption by the UN General Assembly in a dedicated resolution; and (2) a report on the legal aspects of space resources activities, including an assessment of the expected benefits and means of the further development of an international framework for space resources

[8] In particular, during the fifty-fifth session of the Legal Subcommittee, States such as Russia and Belgium, raised concerns about the compatibility of some national legislations with international space law [45].

activities as well as recommendations for further work at the Legal Subcommittee, on this topic [47: 3 and 4].

In this context, Greece submitted at the sixty-first session of the UN COPUOS Legal Subcommittee, namely in 2022, a *Questionnaire related to the discussion of item 15 on potential legal models for the exploration, exploitation, and utilization of space resources*, with the aim to facilitate the exchange of views among member States when discussing said topic [48]. Questions contained in this working paper referred to whether the existing legal framework for outer space activities is sufficient to regulate the use of space resources, or if an *ad hoc* framework is required;[9] whether the issue of the use of space resources should be considered to fall within the scope of Art. II of the OST (i.e., the non-appropriation principle);[10] whether the issue of the use of space resources should be considered with a view to an all-inclusive regulation, or different approaches on a case-by-case basis should be adopted (e.g., in relation to orbits around the earth, to *in situ* resource utilization or to space mining for commercial exploitation); whether a legal framework for the use of space resources should promote an equitable access to said resources, for all States;[11] and ultimately, whether the interaction of this issue with the long-term sustainability of outer space activities should be considered[12] and whether the technical and scientific aspects of the use of space resources should be regarded to be factors that affect the development of a framework.[13]

It is undoubtable that this topic is gaining in importance, and States involved tend to promote approaches and policies adapted to their specific needs and interests. Therefore, debates appear to include such different aspects as "the creation of an international legal framework governing space resources, the interpretation of the OST, the significance of environmental sustainability, the role of the COPUOS

[9]Each of these questions was accompanied by guiding questions (sub-questions), namely: "1.1. If the answer to (a) is in the affirmative, which existing principles and provisions provide for an adequate regulation of the matter?", and "1.2. If the answer to (b) is in the affirmative, what should be the main characteristics of an ad hoc framework?". In addition to that, in case the answer to (b) was affirmative, questions were asked regarding the specific characteristics that said framework should have, [48: 2].

[10]Here, again, guiding questions were included, namely: "2.1 Should Article II of the Outer Space Treaty be subject to a commonly accepted interpretation?"; "2.2 Should Article II of the Outer Space Treaty be amended?" and "2.3 Should there be a review of the concept of celestial bodies, so as to narrow the scope of Article II (e.g. as regards comets and asteroids)?" [48: 2].

[11]In addition to that, the question was asked whether "5. (...) the Moon Agreement provides a suitable basis for the development of a legal regime for the exploration, exploitation and utilization of space resources" and whether "7. (...) the existing international legal regime for mineral exploitation in the Area (international seabed) possibly serves as a model for the exploration, exploitation and utilization of space resources?" [48: 2].

[12]To clarify this question, it was further asked whether "6. In general, the issue should be considered in the light of environmental considerations, in accordance with the letter and spirit of Article IX of the Outer Space Treaty?" [48: 2].

[13]The additional question in this case was: "If the answer is in the affirmative, what is the best way to make the relevant feedback available to the working group on space resource?" [48: 2].

Legal Subcommittee and non-State actors, and the inclusion of developing countries within future space resource initiatives" [45: 16].

In essence, all these developments are a testament to the importance of the debates that currently take place at the international level and the significant initiatives undertaken by Greece at the UN COPUOS are clearly indicative of its determination to participate actively in the said critical areas, as well.

2.2.2 *Participating in Major International Satellite Systems*

In addition to actively participating (with strong interest and commitment) in the principal UN *forum* for space policy, the Greek State ensured its involvement in the major satellite systems. Practically, satellite systems have been exploited in the field of telecommunication and broadcasting for the past sixty years, materializing the objective of Resolution 1721 (XVI) adopted in 1961 [49],[14] as well as Resolution 1802 (XVII) signed in 1962, emphasizing "the importance of international cooperation to achieve effective satellite communications which will be available on a worldwide basis", [50, 42: 282].

Therefore, in years following the adoption of the above-mentioned resolutions, "the UN desire to see the provision of a global satellite system affording non-discriminatory access to all nations was largely met, particularly through the establishment of various organizations both international and national (...)" [42: 284]. In more practical terms, "the first organizations to offer telecommunication facilities, INTELSAT [*and*] INMARSAT (...) were intergovernmental and directed to the public service[15]" [42: 281].

2.2.2.1 Joining the INTELSAT

The *International Telecommunications Satellite Organization* (INTELSAT) was created in 1964 so as to "own and manage a constellation of communications satellites providing international broadcast services" [51], in the form of a

[14] Resolution 1721A (XVI) of 20 December 1961, is regarded by many as the first chapter in the book of space law. More precisely, as it was noted by the US delegate of that time, "When a General Assembly resolution proclaimed principles of international law—as resolution 1721 (XVI) had done—and was adopted unanimously, it represented the law as generally accepted in the international community", [49].

[15] However, it is necessary to note that said organizations "have now become private enterprises, in each case with a residual organization with a supervisory role in relation to the public interest, a role of variable effectiveness. A 'public service' is not the same as a 'service available to the public' where profit is a major objective. Although telecommunications services remain the business conducted by the major companies, profitability rather than public service now appears dominant" [42: 281].

consortium regarded as "a short-term organizational effort which was as experimental and tentative as the technology which it was charged to exploit" [52]. Hence, although initially (i.e., in 1964) set up on an interim basis [42: 287 and 288],[16] it soon took on a permanent form, when the *Agreement relating to the International Telecommunications Satellite Organization 'INTELSAT'* was signed (on 20 August 1971 and entered into force on 12 February 1973 [53]). Greece acceded to this agreement by means of the Legislative Decree n° 1237 of 18 September 1972 [54], containing the full text of the definitive INTELSAT Arrangements (in French), accompanied by a translation in Greek language. Subsequently, the Greek State additionally ratified the *International Telecommunications Satellite Organization 'INTELSAT' Privileges and Immunities Order* which had been adopted on 19 May 1978, on the basis of Law n° 1781 of 31 May 1988 [55]. As a result, said organization enjoyed such privileges and immunities as were necessary for the exercise of its functions and the fulfillment of its objectives.

In the early 2000s, following the commercial and market developments taking place at that point and in order to adapt to the new competitive environment, INTELSAT was privatized and divided into two separate entities. Indeed, the "operational and business elements of INTELSAT were transferred to an interrelated set of commercial companies (known as INTELSAT Ltd) on July 2001, and the new structure operated thereafter. The former parties to the INTELSAT Agreement became shareholders of the new companies in proportion to their previous investment shares (...)". At the same time, the intergovernmental organization remained with a residual character and a new designation (*International Telecommunications Satellite Organization—ITSO*), while its functions were limited to the supervision of the compliance with the basic principles on the part of the new entity [42: 294], following amendment of the initial Agreement (signed on 20 August 1971) which was agreed upon on 17 November 2000.

The Greek State also ratified this second agreement, by means of Law n° 3155 of 17 June 2003 [56]. Consequently, Greece is up to now one of the 149 ITSO member States [57], and thus participates in monitoring the performance of Intelsat's public service obligations while safeguarding the Parties' Common Heritage.

Overall, the setting up of INTELSAT ensured that "the aspirations of the early UN resolutions in favor of the provision of a global telecommunications satellite system with non-discriminatory access for all were in the main fulfilled. The sole obvious exception was the provision of maritime services, which for political and

[16]"Interim INTELSAT existed from 1964 to 1973 and was a consortium of joint-venturers, not a formal international organization. Nonetheless, its structure and principles foreshadowed the definitive arrangements. It took the initial decisions as to the creation of the INTELSAT system and brought it into being. The arrangements consisted of an Interim Agreement (Int. Ag.) between States and a Special Arrangement (Sp. Ag.) between telecommunications operators (PTTs), each nominated one per state by the parties to the Interim Agreement, each designated. (...) Initial membership was by invitation based upon the international ranking of each potential member in volume of international telecommunications using those ITU statistics" [42: 287 and 288].

financial reasons were provided through the separate system INMARSAT" [42: 292].

2.2.2.2 Accession to the INMARSAT

The *International Maritime Satellite Organization* (INMARSAT) was created in 1979 by the *International Maritime Organization* (IMO) so as "to develop a satellite communications network for protecting lives at sea" [58], based on an international agreement signed on 3 September 1976, which came into force on 16 July 1979 [59]. It was created as an inter-governmental body, with the objective to "make provision for the space segment necessary for improving maritime communications, thereby assisting in improving distress and safety of life at sea communications, efficiency and management of ships, maritime public correspondence services and radiodetermination capabilities", and to serve all areas where there is need for maritime communications, acting exclusively for peaceful purposes [59: Art. 3].

The INMARSAT convention was transposed into the internal legal order by means of Law n° 942 dated 5 July 1979 [60], containing the text of the INMARSAT convention (in English) as well as a translation in Greek language. In brief, "the maritime services were available without discrimination (…). Financed by capital contributions from the Signatories and system revenues it operated on a commercial basis (…). Ownership of the space segment was shared by Signatories in proportion to investment share" [42: 298]. In this context, an Annex to the *Operating Agreement on the International Maritime Satellite Organization (INMARSAT)* mentioned that the initial share of Greece was 3,50 out of 101,45, which enabled it to have a (relatively) substantial participation in this endeavor (Greece was the ninth most contributing State, out of a total of 40 countries, with a participation comparable to that of France, Germany, the Netherlands and Canada). As expected, the Geek administration also ratified the *Protocol on the Privileges and Immunities of the International Maritime Satellite Organization (INMARSAT)* [61], by means of Law n° 1783 adopted on 31 May 1988 [62].

Nevertheless, "(a)lthough the convention initially contemplated a maritime service only, the Convention and Operating Agreement were subsequently amended to allow the provision of air and land-mobile communications" [42: 299]. Thus, the decision to develop air communication, adopted by the INMARSAT Assembly in October 1985—i.e, to open maritime satellite systems also for aeronautical communications for the benefit of aircraft of all nations [63], was transposed into the Greek legal order, on the basis of Law n° 1779 of 31 May 1988 [64]. Likewise, the Greek State incorporated into the national legal order the amendment enabling INMARSAT to provide services to land-based vehicles (in addition to ships and aircraft), adopted during the INMARSAT Assembly of 19 January 1989 [64], by means of Law n° 2037 adopted on 9 April 1992 [65].

In 1998, the INMARSAT Assembly took the decision to turn the organization into "a privatized corporate structure, while retaining intergovernmental oversight of certain public service obligations and, in particular, the Global Maritime Distress and

Safety System (GMDSS)" [66]. Hence, the new structure would comprise (i) Inmarsat Ltd., a public limited company forming the commercial arm of the entity and taking on all the commercial activities of INMARSAT (it was completely privatized by the end of 2003) and (ii) the International Mobile Satellite Organization (IMSO), an intergovernmental body established to ensure that INMARSAT continues to meet its public service obligations [42: 299 et seq.]. The amended treaty, as it was approved on 20-24 April 1998 (INMARSAT 12th Assembly) was transposed into the Greek legal order on the basis of Law n° 2875 of 12 December 2000 [67] as well as a certain number of decisions, adopted subsequently, that further improved the practical functioning of the IMSO [42: 300 et seq.].

Participating in international satellites systems for quite a few decades, allowed the Greek State to undoubtedly receive high quality satellite services. At the same time, it evidently realized the importance of having access to reliable space infrastructure, cementing its role in the global space arena.

2.3 Adapting to the Latest Developments in International Space Law (i.e. Soft Law)

Recent decades have been marked by an unprecedented expansion of space activity; the most notable feature in the field of said activity being that a growing number of private companies or public-private joint ventures are getting involved in space matters, and that new space-faring nations are emerging (such developments were undoubtedly aided by the advances in space technology, that led to a clear drop in space robotics prices). At the policy level, specific opinions were promoted, such as that many of the international space law rules are now, to a certain extent, inadequate, given the new political and economic challenges the space sector has to face. As a result, the current framework was criticized for being State-centric and thus poorly adapted to the needs of the new stakeholders (in other words, too rigid to promote the emerging private space sector); following on from that, it was argued that "the golden age of the space treaties, 1967–1979, is now well passed. Many new legal issues have developed since then, but a resurgence of general space treaty-making seems unlikely" [42: 32].

The Greek State kept pace with such changes. It had evidently ratified the principal treaties on space matters, and it also ratified all the major space-related conventions and agreements *lato sensu*, in due time: like, for instance, the *Convention on the Prohibition of Military or Any Other Hostile Use of Environmental Modification Techniques*, signed in 1976 [68], on 23 August 1983 [69]; the instruments relating to the functioning of the ITU, such as the *Constitution and Convention of the International Telecommunication Union* adopting major structural changes and signed in 1992 [70], on 25 September 1998 [3] etc. Now, taking into account the emergence of soft law rules as a means of complementing the existing framework, it furthermore ensures respect for these guiding rules and principles of a new nature to which the leading space-faring countries subscribe.

2.3.1 *Non-binding Rules Negotiated in International Fora: UN Resolutions*

International space law today includes, along with the rules that were established by the relevant space law treaties—and created legally binding obligations upon States—, a growing number of rules and guidelines of a non-binding though important nature, as they establish standards and/or propose solutions to specific technical issues ("notwithstanding the inherent risks that this (potentially) brings of greater 'non-compliance" [28]). These soft law instruments are aimed at providing practical guidance which States and operators are, generally, expected to consider while planning their space policies and/or conducting space operations (i.e., soft law is not binding upon States, unless it finally develops into a customary norm, *via* the repeated practice of States based on consistent *opinion juris*; see Art 38 of the ICJ Statute [71]).

At the global level, the UN system (including the UN Assembly, the UN COPUOS etc.) provides the most reliable platform for communication and negotiation among States, while it allows presentation of the views of both institutional and private actors and stakeholders, as well as of the academia and the civil society. Hence, a number of legal instruments were adopted in this context which—in spite of not being obligatory and binding—had an undeniable moral authority and allowed for the crystallization of new concepts and values that emerged in the context of space activity. Debates on key issues pertaining to space matters and to the legal framework that should apply have, quite often, led to the adoption of resolutions, providing guidance to States and introducing topics for consideration.

By way of illustration, resolutions were enacted to provide guidance on the adoption of national space laws, such as *inter alia* the 2004 Resolution on the Application of the Concept of the 'Launching State' [72]; the 2007 Resolution on the Recommendations on Enhancing the Practice of States and International Intergovernmental Organizations in Registering Space Objects [73]; the 2013 Resolution on Recommendations on National Legislation Relevant to the Peaceful Exploration and Use of Outer Space [74] etc. In truth, "(t)he role of resolutions of the General Assembly of the United Nations in public international law is controversial. Certainly, they are not legislative instruments making law, although they may reflect customary law or contribute towards its development (...). However much one would want to see these statements complied with, it has to be recognized that in practice their autonomy may be limited, and their prescriptions are often ignored and remain controversial" [42: 39 and 40].

Be that as it may, the Greek State made the choice to essentially take into account the UN major resolutions while deciding on space matters, apparently considering them to form "part of the material sources as they clearly indicate the obligations of the State" [75].

More precisely, it is clearly stipulated in the Explanatory Memorandum to the Draft of the Greek space law enacted in 2017 [76] that in adopting national legislation on space activity, account must be taken of the relevant norms of international space law, in particular treaties and agreements adopted within the UN context, creating international rights and legally binding obligations for those States which signed and ratified them; that is, on the basis, for Greece, of Art. 28 para. 1 of the Greek constitution [76: 1]. Nevertheless, to avoid any ambiguity and to guarantee the uniform application of international space law, it is furthermore highlighted that apart from binding legislation, account must also be taken of a series of decisions taken by the UN Security Council as well as of other legal instruments—importantly, with a specific mention to guidelines, recommendations, principles, rules and regulations—, regulating space matters in greater detail [76: 1].

Interestingly, the Greek legislator did not precisely enumerate those resolutions—or soft-law documents—to take into consideration, leaving national competent authorities with the choice of the most appropriate instruments. However, an explicit reference was made to the fact that the usefulness of such soft-law guidelines shows the importance of actively participating in *fora* and coordinating bodies aimed at developing—both hard and soft-law—rules regulating space activity issues, either at the international (mention was made to the UN COPUOS) or at the regional level (e.g., European Space Agency) [76: 1].

2.3.2 Guidelines and Rules Established by Space Entities and the Space Industry

In addition to the critical legal instruments adopted at the UN level, like resolutions, a wide range of soft-law rules and guidelines have also been developed to address space activity issues at the initiative of space actors such as operators, companies (governed either by private or public law) and national space agencies. Most of those instruments may be regarded as bottom-up approaches—i.e., they consist in the "contributions of international institutions and agencies that engage in or are otherwise involved in space both at the policy and law-making levels" [42: 19]—, while they are aimed at tackling specific challenges in the field, and at facilitating the conduct of space operations to the benefit of all actors involved; such as the *IADC Guidelines*, which were released by the *Inter-Agency Space Debris Coordination Committee (IADC)*.

The IADC may be defined as an international forum of governmental bodies for the coordination of activities related to the issues of man-made and natural debris in space [77]. Hence, its objective is to "exchange information on space debris research activities between member space agencies, to facilitate opportunities for cooperation in space debris research, to review the progress of ongoing cooperative activities, and to identify debris mitigation options" [77].

In this context, the IADC released (in 2002) a set of voluntary guidelines, to provide space operators and competent authorities with specific information and rules for the design, launch and operation of space objects. The document, entitled *IADC Guidelines*, included a series of standards and practices developed by all its member organizations, and was aimed at reducing the creation of space debris during space routine operations; its overall purpose was to gather the best expertise available, to minimize the potential for accidental on-orbit breakups, to regulate the disposal of spacecraft post-mission, to prevent on-orbit collisions and to avoid intentional destruction as well as other harmful activities [77]. Given the critical importance of this challenge, the *IADC guidelines* were built upon, and paved the way for the *Space Debris Mitigation Guidelines of the Committee on the Peaceful Uses of Outer Space* which were later endorsed by the UN General Assembly [78].

In practice, the initiative is said to have had a limited impact: strong commitments were made only by a few space-faring nations, such as the Russian Federation and the United States. Nevertheless, as not all States have made the same level of commitments or have a common practice for mitigating and removing space debris [79], the *IADC guidelines* undoubtedly shed some light on the appropriate measures, despite being "limited in scope, clarity and legal status" [80].

Focusing on Greece, it could be argued that the State participated—indirectly—in the development of the guidelines, given that the IADC members include the European Space Agency (ESA) with which Greece is in close collaboration since the early 1990s, becoming a full member State in 2005. Therefore, given the importance of the subject addressed in the *IADC guidelines*—i.e., there is no longer any doubt that, unless strong measures are taken to reverse the trend, the number of space debris will continue to grow, strongly increasing the risk of collision—, and although the Explanatory Memorandum to the Draft of the Greek space law contains no specific mention of the *IADC guidelines*, it is clear that they must apply; particularly considering that explicit reference is made to other comparable instruments, such as the ESA (European Space Agency) "Space Debris Mitigation Compliance Verification Guidelines" [76: 3].

In any case, the Explanatory Memorandum to the Greek space law underlines that when participating in space activity (namely, including by adopting laws on such matters), the administration must necessarily consider *any* officially adopted rules and guidelines applying to the conduct of space operations; the reason which is mentioned is—*inter alia*—that it shall bear, as launching State, international responsibility towards third parties for damage caused by its space objects in outer space [76: 1]. Hence, it will only benefit from taking into account the standards and guidelines already established by experienced space actors in this domain. In addition to that, the Greek State clarified (see e.g., the *Compendium of standards adopted by States and international organizations to mitigate space debris* in 2019 and 2021 [81]) that, to address the space debris mitigation problem, the Ministers of Digital Policy, Telecommunications and Media, and Environment and Energy, may issue "a joint ministerial decision (...) which will define the content of the reports, the requirements and the technical standards, which the activity of the organization must comply with". In parallel, it was highlighted that: "until this decision is adopted, relevant national, European and international standards, as well as relevant good practices, are applied" [81].

Thus, it is apparent that the Greek State is showing the highest level of engagement also in the implementation of commonly agreed standards and practices adopted in the context of soft-law initiatives, to guarantee the safe conduct of all space activities under its jurisdiction. In this sense, it made clear its readiness to abide by both hard and soft-law rules and laws in effect, giving the highest priority to conducting space activity in a safe and sustainable manner.

In conclusion, it is the increased involvement of the Greek operators and industry in space activities that triggered the constantly growing participation of the Greek State in space policy-making, through the development of specific views; such as *via* the formulation of proposals to be submitted for debate in international space law *fora*, and observations and suggestions that were made to the competent authorities and bodies, or through the co-decision of relevant guidelines and common technical standards. Indeed, practially speaking, such a policy-related engagement was made possible by the Greek active and ambitious participation in several scientific and research space programs—as well as inter-agency agreements—mainly operated by the European Space Agency (ESA) and the EU, that will be presented below.

References

1. ITU. *International Telegraph Convention/Convention télégraphique internationale de Paris (1865) et Règlement de service international (1865).* Accessed from https://search.itu.int/history/HistoryDigitalCollectionDocLibrary/4.1.43.fr.200.pdf
2. UN COPUOS STSC. (2021). *Statements to the 945th meeting of the Scientific and Technical Subcommittee, Monday, 26 April 2021 - Item 3: General exchange of views and introduction of reports submitted on national activities (Greece).* Accessed from https://www.unoosa.org/oosa/en/ourwork/copuos/stsc/2021/statements.html
3. International Telecommunication Union (ITU). (2024). *List of member states.* Accessed July 2024, from https://www.itu.int/online/mm/scripts/gensel8
4. International Telecommunication Union (ITU). (2024). *Overview of ITU's History (1).* Accessed July 2024, from https://www.itu.int/en/history/Pages/ITUsHistory.aspx
5. International Telecommunication Union (ITU). (1968). *ITU and space radiocommunication.* ITU. Accessed from https://search.itu.int/history/HistoryDigitalCollectionDocLibrary/12.101.73.en.100.pdf
6. UN. *Declaration of legal principles governing the activities of states in the exploration and use of outer space.* GA Res 1962 (XVIII), UNGAOR, 18th Session.
7. CVCE.eu. *Treaty banning nuclear weapons tests in the atmosphere, in outer space and under water. 5 August 1963. UNTS 480, p. 43 (entered into force on 10 October 1963).* Accessed from https://treaties.un.org/pages/showDetails.aspx?objid=08000002801313d9 (*Nuclear Test Ban Treaty*).
8. United Nations Office for Outer Space Affairs. *Treaty on principles governing the activities of states in the exploration and use of outer space, including the moon and other celestial bodies.* GA Res 2222 (XXI), signed on 27 January 1967 (entered into force, 10 October 1967).

Accessed from https://www.unoosa.org/oosa/oosadoc/data/resolutions/1966/general_assembly_21st_session/res_2222_xxi.html (*Outer Space Treaty; OST*).
9. Hofmann, M., & Bergamasco, F. (2020). Space resources activities from the perspective of sustainability: Legal aspects. *Global Sustainability, 3*. Accessed from https://www.cambridge.org/core/journals/global-sustainability/article/space-resources-activities-from-the-perspective-of-sustainability-legal-aspects/DF153F4A77970AC9E12444EC2B001F8A
10. United Nations Office for Outer Space Affairs. *Agreement on the rescue of Astronauts, the return of Astronauts and the return of objects launched into outer space (Rescue Agreement), signed on April 22nd, 1968 and entered into force in December 3rd, 1968*. Accessed from https://www.unoosa.org/oosa/en/ourwork/spacelaw/treaties/introrescueagreement.html
11. United Nations Office for Outer Space Affairs. *Convention on international liability for damage caused by space objects (liability convention), signed on March 29th, 1972 and entered into force in September 1st, 1972*. Accessed from https://www.unoosa.org/oosa/en/ourwork/spacelaw/treaties/introliability-convention.html
12. United Nations Office for Outer Space Affairs. *Convention on registration of objects launched into outer space (registration convention), signed in January 14th, 1975 and entered into force in September 15th, 1976*. Accessed from https://www.unoosa.org/oosa/en/ourwork/spacelaw/treaties/introregistration-convention.html
13. United Nations Office for Outer Space Affairs. *Agreement governing the activities of states on the moon and other celestial bodies, 18 December 1979 (entered into force 11 July 1984), UNTS 1363*, 3. Accessed from https://www.unoosa.org/oosa/en/ourwork/spacelaw/treaties/intromoon-agreement.html (*Moon Agreement*).
14. United Nations Treaty Collection (*UNTC*). (n.d.). *OST - Ratifications*. Accessed July 2024, from https://treaties.un.org/pages/showdetails.aspx?objid=0800000280128cbd
15. *Legislative Decree n° 670/1970 on the ratification of the Treaty on principles governing the activities of states in the exploration and use of outer space, including the moon and other celestial bodies signed on 27.1.67*. Greek Government Gazette, Vol. A '208.
16. Cambridge Advanced Learner's Dictionary and Thesaurus. (n.d.). *Meaning of "province" and "someone's province"*. Accessed July 2024, from https://dictionary.cambridge.org/dictionary/english/province
17. MacMillan Dictionary. (n.d.). *Meaning of "province"*. Accessed July 2024, from https://www.macmillandictionary.com/dictionary/british/province
18. Walsh, J. W. (1901). The true meaning of the term "jurisdiction". *The American Law Register (1898-1907), 49*(6), 346–356. https://doi.org/10.2307/3306463
19. Vienna Convention on the Law of Treaties. 23 May 1969. (entered into force 27 January 1980). *Treaty Series, 1155*, 331. (*Vienna Convention*).
20. Kuner, C. B. (1991). The interpretation of multilingual treaties: Comparison of texts versus the presumption of similar meaning. *International and Comparative Law Quarterly, 40*(4), 953–964. https://doi.org/10.1093/iclqaj/40.4.953
21. United Nations Treaty Collection (*UNTC*). (n.d.). *Rescue agreement - Ratifications*. Accessed July 2024, from https://treaties.un.org/Pages/showDetails.aspx?objid=080000028012504f
22. *Legislative Decree n° 189/1974, on the Ratification of the Rescue Agreement*. Published in the Greek Government Gazette A '350 (pp. 2572a et seq).
23. United Nations Treaty Collection (UNTC). (n.d.). *Liability convention – Ratifications*. Accessed July 2024, from https://treaties.un.org/pages/showdetails.aspx?objid=08000002801098c7
24. *Law n° 563/1977, on the ratification of the liability convention*. Published in the Greek Government Gazette A '75 (pp. 595 et seq.).
25. United Nations Treaty Collection (*UNTC*). (n.d.). *Registration convention – Ratifications*. Accessed July 2024, from https://treaties.un.org/Pages/ViewDetailsIII.aspx?src=TREATY&mtdsg_no=XXIV-1&chapter=24&Temp=mtdsg3&clang=_en
26. *Law n° 3131/2003 on the Ratification of the Registration convention*. Published in the Greek Government Gazette A '79 (pp. 1279 et seq.)

27. United Nations Treaty Collection (*UNTC*). (n.d.). *Moon Agreement – Ratifications*. Accessed July 2024, from https://treaties.un.org/Pages/ViewDetails.aspx?src=IND&mtdsg_no=XXIV-2&chapter=24&clang=_en
28. Jakhu, R. S., & Freeland, S. R. (2016). The relationship between the outer space treaty and customary international law. In *Proceedings of the 67th International Astronautical Congress (IAC 2016): Making space accessible and affordable to all countries* (pp. 11648–11659). International Astronautical Federation, IAF. Accessed from https://iafastro.directory/iac/archive/browse/IAC-16/E7/2/32294/
29. Hellenic Republic – Ministry of Foreign Affairs. (2023). *Greece in international Organizations, UN*. Accessed July 2024, from https://www.mfa.gr/en/foreign-policy/greece-in-international-organizations/un.html
30. United Nations, Dag Hammarskjöld Library. (n.d.). *Founding member states*. Accessed from https://research.un.org/en/unmembers/founders accessed July 2024
31. United Nations. *Res 1348 (XIII) Question of the peaceful use of outer space*. UNGAOR 13th Session, Suppl. no. 18, UN Doc *A_RES_1348(XIII)-EN. 13 December 1958*. Accessed from https://digitallibrary.un.org/record/206866?v=pdf
32. United Nations. *Res 1472 (XIV) International co-operation in the peaceful uses of outer space*. UNGAOR, 14th Session, UN Doc *A_RES*_1472 (XIV). 12 December 1959. Accessed from https://www.unoosa.org/pdf/gares/ARES_14_1472E.pdf
33. UNOOSA. (n.d.). *History*. Accessed July 2024, from https://www.unoosa.org/oosa/en/aboutus/history/index.html
34. United Nations. *Res 35/16 enlargement of the committee on the peaceful uses of outer space*. UNGAOR, 35th Session, UN Doc *A_RES*_35/16. 3 November 1980. Accessed from https://digitallibrary.un.org/record/17219?ln=en&v=pdf
35. UNCOPUOS LSC. (2011). *Review of the legal aspects of the space debris mitigation guidelines of the committee on the peaceful uses of outer space, with a view to transforming the guidelines into a set of principles to be adopted by the general assembly. Working paper submitted by the Czech Republic*. UN COPUOS 50th Session. UN Doc A/AC.105/C.2/L.283. Accessed from https://www.unoosa.org/oosa/oosadoc/data/documents/2011/aac.105c.2l/aac.105c.2l.283_0.html
36. UNCOPUOS LSC. (2002). *Proposal to recommend that the Legal Subcommittee at its forty-second session review the issue of the appropriateness and desirability of drafting a universal comprehensive convention on international space law as a sub-item of the item of its agenda entitled "Status and application of the five United Nations treaties on outer space". Working paper submitted by China, Greece and the Russian Federation*. UN COPUOS LSC 41st session, UN Doc A/AC.105/C.2/L.236. Accessed from https://digitallibrary.un.org/record/477619?ln=ar&v=pdf
37. UNCOPUOS. (2003, April 10). *Report of the Legal Subcommittee on the work of its forty-second session, held in Vienna from 24 March to 4 April 2003 - Forty-sixth session*. UN Doc A/AC.105/805. Accessed from https://www.unoosa.org/pdf/reports/ac105/AC105_805E.pdf
38. UNCOPUOS. (2007, May 2). *Report of the Legal Subcommittee on its forty-sixth session, held in Vienna from 26 March to 5 April 2007 - Fiftieth session*. UN DOC A/AC.105/891. Accessed from https://www.unoosa.org/pdf/reports/ac105/AC105_891E.pdf
39. UN Secretariat COPUOS. (2008). *Information furnished in conformity with the Convention on Registration of Objects launched into outer space note verbale dated 14 May 2008 from the permanent mission of Greece to the United Nations (Vienna) addressed to the Secretary-General. 15 May 2008*. UN DOC ST/SG/SER.E/INF.20. Accessed from https://digitallibrary.un.org/record/630856?ln=en&v=pdf
40. UN COPUOS LSC. (2017, March 28). *Matters relating to the definition and delimitation of outer space: replies of Greece*. UN Doc A/AC.105/C.2/2017/CRP.16. Accessed from https://www.unoosa.org/res/oosadoc/data/documents/2017/aac_105c_22017crp/aac_105c_22017crp_16_0_html/AC105_C2_2017_CRP16E.pdf

41. UN COPUOS LSC. (2019, March 28). *Matters relating to the definition and delimitation of outer space: replies of Greece. National legislation and practice relating to the definition and delimitation of outer space.* UN Doc A/AC.105/C.2/2019/CRP.6. Accessed from https://www.unoosa.org/res/oosadoc/data/documents/2019/aac_105c_22019crp/aac_105c_22019crp_6_0_html/AC105_C2_2019_CRP06E.pdf
42. Lyall, F., & Larsen, P. B. (2018). *Space law: A treatise* (2nd ed.). Routledge. https://doi.org/10.4324/9781315242712
43. U.S. Commercial Space Launch Competitiveness Act. *Public Law 114–90—Nov. 25, 2015.* Accessed from https://www.congress.gov/114/plaws/publ90/PLAW-114publ90.pdf
44. Luxembourg Space Agency. *Luxembourg Law of July 20th 2017 on the exploration and use of space resources.* Accessed from https://space-agency.public.lu/en/agency/legal-framework/law_space_resources_english_translation.html
45. Campbell, A. (2021). *Space resource discussions within the United Nations committee on the peaceful uses of outer space legal subcommittee: The past, present and future.* Institute for Earth and Space Exploration White Papers. 1. Accessed from https://ir.lib.uwo.ca/iesewp/1
46. UN COPUOS LSC. (2019, March 4). *Proposal for the establishment of a working group for the development of an international regime for the utilization and exploitation of space resources Working paper by Belgium and Greece.* UN DOC A/AC.105/C.2/L.311. Accessed from https://www.unoosa.org/res/oosadoc/data/documents/2019/aac_105c_2l/aac_105c_2l_311_0_html/AC105_C2_L311E.pdf
47. UN COPUOS LSC. (2021, June 8). *The establishment of a working group on potential legal models for activities in exploration, exploitation and utilization of space resources - Proposal submitted by Austria, Belgium, Czech Republic, Finland, Germany, Greece, Poland, Portugal, Romania, Slovakia and Spain.* UN DOC A/AC.105/C.2/2021/CRP.22. Accessed from https://www.unoosa.org/res/oosadoc/data/documents/2021/aac_105c_22021crp/aac_105c_22021crp_22_0_html/AC105_C2_2021_CRP22E.pdf
48. UN COPUOS LSC. (2022, March 28). *Questionnaire related to the discussion of item 15 on potential legal models for the exploration, exploitation, and utilization of space resources - Working Paper submitted by Greece.* UN DOC A/AC.105/C.2/2022/CRP.13*. Accessed from https://www.unoosa.org/res/oosadoc/data/documents/2022/aac_105c_22022crp/aac_105c_22022crp_13_0_html/AAC105_C2_2022_CRP13E.pdf
49. Cheng, B. (1997). United Nations resolutions on outer space: 'Instant' International Customary Law? In *Studies in international space law* (online edn). Oxford Academic https://doi.org/10.1093/ACPROF:OSO/9780198257301.003.0007
50. UNOOSA. *Res 1802 (XVII) international co-operation in the peaceful uses of outer space.* UNGAOR, 17th Session (14 December 1962). Accessed from https://www.unoosa.org/pdf/gares/ARES_17_1802E.pdf
51. INTELSAT. (n.d.). *Intelsat history.* Accessed July 2024, from https://www.intelsat.com/intelsat-history/
52. Levy, S. A. (1975). INTELSAT: Technology, politics and the transformation of a regime. *International Organization, 29*(3), 655–680. Accessed from http://www.jstor.org/stable/2706345
53. UNTS. *Agreement relating to the International telecommunications satellite organization 'INTELSAT', signed on 20 August 1971 (entered into force 12 February 1973)* (Vol. 1220, p. 21). UNTS. Accessed from https://treaties.un.org/Pages/showDetails.aspx?objid=08000002800e8e08&clang=_en
54. *Legislative Decree n° 1237/1972 on the ratification of the agreement relating to the International Telecommunications Satellite Organization 'INTELSAT'.* Greek Government Gazette A' 191/31.10.1972, p. 1645 et seq.
55. *Law n° 1781/1988 on the ratification of the protocol pertaining to the privileges, exemptions and immunities of the International Telecommunications Satellite Organization (INTELSAT).* Greek Government Gazette A' 115/03.06.1988, p. 2291. et seq.

56. *Law n° 3155/2003 on the ratification of the agreement on the International Telecommunications Satellite Organization (ISTO) and amendment of article 23 of the operating agreement of the organization.* Greek Government Gazette A' 155/19.06.2003, p. 2385. et seq.
57. ITSO. (n.d.). *About us*. Accessed July 20204, from https://itso.int/about-us/
58. INMARSAT. (n.d.). *About, who we are*. Accessed from https://www.inmarsat.com/en/about/who-we-are.html accessed July 2024
59. IMO. *Convention on the International Maritime Satellite Organization (INMARSAT), signed on 3 September 1976, entered into force 16 July 1979*, UNTS 1143, 105.
60. *Law n° 942/1979 on the ratification of the agreements pertaining to the International Maritime Satellite Organization (INMARSAT).* Greek Government Gazette A' 155/10.07.1979, p. 1605 et seq.
61. University of Oslo. *Protocol on the privileges and immunities of the International Maritime Satellite Organization (INMARSAT), signed in 1 December 1981, entered into force in July 3rd, 1983.*
62. *Law n° 1783/1988 on the ratification of the protocol on the privileges and immunities of the International Maritime Satellite Organization (INMARSAT).* Greek Government Gazette A' 118/07.07.1988, p. 2339 et seq.
63. Von Noorden, W. D., & Dann, P. (1989). Land mobile satellite communications: A further development in international space law (Part I). *Journal of Space Law, 17*(1), 1.
64. *Law n° 1779/1988 on the ratification of the amendments to the convention and operating agreement of the International Maritime Satellite Organization (INMARSAT) as ratified by Law n° 942/1979.* Greek Government Gazette A' 112/01.07.1988, p. 2273 et seq.
65. *Law n° 2037/1992 on the Ratification of the amendments to the Convention and Operating Agreement of the International Maritime Satellite Organization (INMARSAT) as ratified by Law n° 942/1979 and Law n° 1779/1988.* Greek Government Gazette A' 61/13.04.1992, p. 1331 et seq.
66. UNCOPUOS LSC. (2008, April 07). *Information on the activities of international intergovernmental and non-governmental organizations relating to space law. Note by the Secretariat.* Forty-seventh session. UN DOC A/AC.105/C.2/2008/CRP.13. Accessed from https://www.unoosa.org/pdf/limited/c2/AC105_C2_2008_CRP13E.pdf
67. *Law n° 2875/2000 on the ratification of the amendments to the convention of the International Maritime Satellite Organization (INMARSAT) and of the amendments to its Operating Agreement.* Greek Government Gazette A' 287/29.12.2000, p. 4117. et seq.
68. UNTS. *Convention on the prohibition of military or any other hostile use of environmental modification techniques*, 10 December 1976, UNTS 1108, p. 151(entered into force on 5 October 1978).
69. United Nations Treaty Collection (UNTC). (n.d.). *Convention on the prohibition of military or any other hostile use of environmental modification techniques - Ratifications*. Accessed July 2024, from https://treaties.un.org/Pages/ViewDetails.aspx?src=IND&mtdsg_no=XXVI-1&chapter=26&clang=_en
70. UNTS. *Constitution and convention of the International Telecommunication Union*, 22 December 1992, UNTS 1825, 1826 (entered into force on 1 July 1994).
71. International Court of Justice (ICJ). *Statute of the International Court of Justice*. Accessed July 2024, from https://www.icj-cij.org/statute
72. General Assembly Resolution. *59/115 Application of the concept of the 'launching state'*. UNGA 59th Session, UN Doc A/RES/59/115 (25 January 2005), [Resolution on the Concept of 'Launching State'].
73. General Assembly Resolution. *62/101 Recommendations on enhancing the practice of states and international intergovernmental organizations in registering space objects*. UNGA 62nd Session, UN Doc A/RES/62/101 (10 January 2008), (Resolution on Enhancing Registration).
74. General Assembly Resolution. *68/74 Recommendation on National Legislation relevant to the peaceful exploration and use of outer space*. UNGA 68th Session, UN Doc A/RES/68/74 (16 December 2013).

75. Alsharqawi, A., Hamdan, A. B., & Moh'd Abu Anzeh. (2021). The role of general assembly resolutions to the development of international law. *Journal of Legal, Ethical and Regulatory, 24*(2) Accessed from https://www.abacademies.org/articles/the-role-of-general-assembly-resolutions-to-the-development-of-international-law-10426.html
76. *Explanatory memorandum to the draft law entitled authorization of space activities - Registration in the National Registry of Space Objects - Establishment of the Hellenic Space Agency and other provisions.* Accessed from https://www.hellenicparliament.gr/Nomothetiko-Ergo/Anazitisi-Nomothetikou-Ergou?law_id=ca706be4-0ccc-4dbf-a8ce-a83b01095b01 (in Greek).
77. UNCOPUOS STSC. (2002). *Inter-Agency Space Debris Coordination Committee space debris mitigation guidelines.* UN Doc A/AC.105/C.1/L.260 (29 November 2002). Accessed from https://www.unoosa.org/pdf/limited/c1/AC105_C1_L260E.pdf
78. *Res 62/217 International cooperation in the peaceful uses of outer space.* UNGA Sixty-second session. UN Doc A/RES/62/217* (1 February 2008). Accessed from https://www.unoosa.org/pdf/gares/ARES_62_217E.pdf
79. Meegan, M. (2021). The pressing need for legal certainty for space operators to mitigate and removal space debris. In T. Flohrer, S. Lemmens, & F. Schmitz (Eds.), *Proceedings of 8th European Conference on Space Debris (virtual), Darmstadt, Germany, 20–23 April 2021.* ESA Space Debris Office.
80. Hobe, S. (2009). Legal analysis of the UN Space Debris mitigation guidelines (presentation). In *International Interdisciplinary Congress on Space Debris, 7–9 May 2009.* McGill University. Accessed from https://www.mcgill.ca/iasl/files/iasl/Session_2_Stephan_Hobe.pdf
81. UNOOSA. *Compendium of space debris mitigation standards adopted by states and international organizations.* UNOOSA. Accessed from https://www.unoosa.org/oosa/en/ourwork/topics/space-debris/compendium.html accessed July 2024

Chapter 3
Impetus Gained from the Involvement in Regional (EU) Space Initiatives

Having joined the major international space law organizations and *fora*, Greece rapidly started gaining expertise in space matters also by participating in the numerous regional –i.e., European—space initiatives and research programs. In brief, given the significant progress in particular fields of space and space-related activities, such as space communications or the development and use of satellite technology for meteorology, "intergovernmental cooperation appeared at least to ensure a more viable and economically feasible system in satisfying the (...) needs of the European people through the application of a common space system" [1: 190]. As a result, the idea rapidly gained acceptance that specific organizational structures for managing such kind of space assets and space services must be created; i.e., principally "among countries situated in the same geographic area [*having comparable*] motivations of a political, economic, linguistic or cultural nature" [1: 158].

At European level, it is the *European Space Agency (ESA)* which appears to have had the key role in promoting cooperation among European States in the fields of space research and technology, with a view to their being used for scientific purposes and for operational space applications systems, as established in its convention adopted in 1975 [2: Art. III]. However, one must note that a number of European initiatives for space-related cooperation had already seen the light of day as early as in the 60s,[1] like the *European Launcher Development Organization* (ELDO) and the

[1] In brief, (i) the *European Preparatory Commission for Space Research* (COPERS—*Commission Préparatoire Européenne de Recherches Spatiales*) was founded in 1961 with the aim to "prepare for the establishment of an international organization capable of planning and implementing programmes of common interest for Europe's entry into space", which resulted in the creation of the ESRO; while (ii) the *European Conferences on Satellite Communications* (CETS in French/*ECSC* in English) created in 1963 allowed for the "setup of a European communication programme that materialized with the first European telecommunications satellite ... developed by ESRO and launched by ESA", and the drawing up of a common European policy together with the other member countries of the *European Conference of Postal and Telecommunications Administrations* (CEPT, created in 1959). Finally, the first (iii) *European Space Conference* (*ESC*) was held in 1966 so as to coordinate the activities of the ELDO and the ESRO –both created in 1964–, and finally

A. Koskina, M. Plionis, *Greek Space Law*, Space and Society,
https://doi.org/10.1007/978-3-031-72813-6_3

European Space Research Organization (ESRO)—both established in 1964–[2] as well as several inter-State and inter-agency collaborations in this field. Be that as it may, the ESA convention which was signed by the representatives of the Member States of the two principal European space organizations mentioned above "with the intention that these two organizations should be replaced[3]", on 30 May 1975, gave a crucial impetus to European space-related activity (thus, the *Convention for the Establishment of a European Space Agency* was signed in 1975, and entered into force in 1980 [2]).

Interestingly, although the Greek State did participate (in 1965) in space-related initiatives, like the *European Conference on Satellite Communications* (CETS) which aimed at "studying the potential of space telecommunications and encouraging other organizations and associations active" [5: 305] or the *European Conference of Postal and Telecommunications Administrations* (CEPT), it did not really become a member of either the ELDO or the ESRO [6: 294]. As a matter of fact, it only got involved at the European level by participating in the first European organizational structures such as the EUTELSAT and EUMETSAT, established as intergovernmental organizations aimed at managing the space systems and services; and then by getting involved in core aspects of the functioning of the ESA (Sect. 3.1). Such involvement allowed Greece to acquire knowledge and experience on space matters. As a result, it became able to develop and learn within the EU environment, to finally create a domestic space law framework and a promising space sector (Sect. 3.2).

The purpose of these two sections is to provide general information on the steps taken by Greece to establish its voice and participation in space matters at the regional level, and then to develop its own tools and actions in this field of activity. Thus, in line with this reasoning, the key points in the country's path presented below will follow a chronological order (i.e., based on the order of the Greek State' accession to said space programs, organizations and initiatives).

---

reached the agreement, in 1972, that a single *European Space Agency* (ESA) should be created (the final agreement was submitted to the ministerial meeting of the ESC in 1975) [3, 4].

[2]The ELDO (*European Space Vehicle Launcher Development Organization*) was created in 1964 to establish a satellite launch vehicle for Europe; the ESRO (*European Space Research Organization*) was also founded in 1964—by 10 European States- to allow the joint pursue of scientific research in the field of space. The two entities were merged together to form ESA in 1975 [4].

[3]Namely, the *European Launcher Development Organization* (ELDO) and the *European Space Research Organization* (ESRO) [1: 159].

3.1 Joining Management Structures & the ESA Cooperation Framework

From the perspective of the Greek State, its space-related activity—carried out at the European level, and in the chronological order that such steps were decided—shall be divided (for the purposes of this study i.e., to provide an overall image of the measures taken at the national level in this regard), into two broad categories: involvement in management structures, such as the EUMETSAT and EUTELSAT, then involvement in the core framework of the European Space Agency (ESA), which is pursuing a much wider objective and policy; the aim being in each case to highlight the efforts made by the country, in a logical order.

In general terms,—and putting aside its involvement in the first European entities dealing with specific (i.e., precisely defined) space-related matters[4]—, Greece really joined the European effort to establish a strong cooperation among States in space activity, by participating in European organizational structures in the late 80s (as in the case of its involvement in international space-related initiatives); from this viewpoint, said involvement dates back to almost four decades. Following on from that, the Greek State soon acquired expertise and gained an interest in also participating in ESA core activities. Hence, it further applied for a full ESA membership, so as to enhance its engagement with other member States and its commitment to the implementation of the ESA perspective.

3.1.1 *Participating in Management (or Organizational) Structures for Space Services*

European operational space-related services were first successfully established in the fields of telecommunication and meteorology [7: 8] and it soon became clear that these "operational space systems and their management, although promoted by the structure and purposes of the European Space Agency, were bound to be organized in a later phase within a separate framework" [1: 176]. Hence, the *European Telecommunications Satellite Organization* (EUTELSAT) was founded in 1977 with the objective "to operate the first generation of communications satellites ordered by the European Space Agency (ESA)" [8]. Similarly, a *European Organization for the Exploitation of Meteorological Satellites* (EUMETSAT) was created in 1986 as an intergovernmental organization to operate meteorological satellites and to serve the needs of its member States [9], such as by taking over the Meteosat geostationary satellite programme that had been started in the 1970s as an ESA programme [10].

[4] That is, the CETS (*European Conferences on Satellite Communications*) and the CEPT (*European Conference of Postal and Telecommunications Administrations*).

Since the two above-mentioned European organizational structures were created as Inter-Governmental Agencies (IGO), the Greek State's participation in said organizations was established following acceptance of their founding instruments, in accordance with national and international law.

3.1.1.1 Accession to the EUTELSAT: Greek Law n° 1713/1987

In 1973, the ESRO initiated a European telecommunications satellite program, which started as a preoperational program named Orbital Test Satellite (OTS) [1: 176]; OTS-1 was developed by ESRO and launched by ESA in 1977 [11]. As a result, in order to integrate this new satellite system within the existing European communications infrastructure, the *European Conference of Postal and Telecommunications Administrations* (CEPT) had to cooperate with the ESRO and ESA. In this context, the decision was also taken to create a specific organization representing the interests of the CEPT members and to operate the ECS system (namely, "The European Communications Satellite (ECS) was the first European regional satellite communications system, operated by Eutelsat, with ESA providing the first-generation space segment" [12]) on a commercial basis, which led to the creation of Interim EUTELSAT (soon reshaped into EUTELSAT).

The *European Telecommunications Satellite Organization* (hereafter, EUTELSAT) was created in 1977, as "part of the intention of Europe to develop and maintain a space industry east of the Atlantic, the intention that helped produce INTELSAT", since the "ESA members sought to foster their domestic space industries which had not been as successful in gaining INTELSAT contracts as had been hoped" [5: 305 and 306]. The *Agreement on the Constitution of a Provisional European Telecommunications Satellite Organization* (Interim EUTELSAT) [5: 306] created first an association of 17 European countries [13] or authorized private entities, "to ensure the establishment, operation and maintenance of the space segment of satellite telecommunications systems" [1: 177]. Following on from that, the constitutive text of EUTELSAT—i.e., the Convention and an Operating Agreement—was opened for signature in July 1982, and entered into force on 1 September 1985 [14].

The EUTELSAT Convention was an international intergovernmental agreement concluded between States, and open to all European States members of the ITU (while, "the Operating Agreement was concluded between the 'Signatories', being either the relevant governments themselves or their authorized telecommunication entities" [1, at 179]: indeed, the EUTELSAT Convention mentioned that "Each Party shall designate a public or private telecommunications entity subject to its jurisdiction to sign the Operating Agreement, unless such Party itself signs the Operating Agreement" [14: Art. II (b)]). On this basis, the Greek State signed the EUTELSAT Convention on 26 August 1987; its accession came into force on 25 October 1987 [15], following adoption of law n° 1713/1987 entitled "*Ratification of the Convention and Agreement pertaining to the creation and functioning of the European Telecommunications Satellite Organization (EUTELSAT), together with its Annexes and Protocols"* [16] (the Greek law contained the French version of said Convention and Annexes, together with their translation in Greek).

Understandably, Greece additionally ratified the Protocol on the Privileges and Immunities [17] by means of law n° 2280/1995 titled "*Ratification of the Protocol on the Privileges and Immunities of the European Telecommunications Satellite Organization (EUTELSAT)*" [18].

As a result, the organizational arrangements were binding for the Greek State; such as, *inter alia*, Art. IV establishing the legal personality and full capacity of EUTELSAT necessary for the exercise of its functions and the achievement of its purposes (on this basis, EUTELSAT could exercise its functions in an effective manner, as a commercial organization, operating for the benefit of all countries involved [1: 186]); Art. VI mentioning that EUTELSAT would be led by the Assembly of Parties, the Board of Signatories and an executive organ, headed by a Director General (their competences were developed in subsequent articles of the EUTELSAT Convention, namely Art. VII to Art. XIII); Art. XV laying down the rights and obligations of the Parties and Signatories; Art. XVII specifying all the EUTELSAT privileges, exemptions and immunities etc. Furthermore, it was clearly stipulated that "Each Signatory shall have a financial interest in EUTELSAT in proportion to its investment share and this shall correspond to its percentage of all utilization of the EUTELSAT Space Segment by all Signatories as determined under the Operating Agreement" [14: Art. V (c)]. Accordingly, point (1) of Annex B mentioned, for the Greek State, an initial investment share of 3.19%.

In the following years, EUTELSAT had to face growing competition; therefore, member States took the decision to restructure it.[5] Hence, *on the one hand*, all the assets, operational activities, and related liabilities and commitments of the intergovernmental organization EUTELSAT were transferred to Eutelsat S.A. [13]. For this purpose, any amendments to the original EUTELSAT Convention were agreed by the Assembly of Parties in May 1999; likewise, all the new arrangements were applied on an accelerated (provisional) basis from 2 July 2001 [13] and the amended Convention entered into force on 28 November 2002. The Greek State ratified these amendments, according to law n° 2962/2001 entitled "*Ratification of the amended Convention on the establishment of the European Telecommunications Satellite Organization (EUTELSAT)*" [19]. The new Eutelsat S.A. had the form of a commercial entity—namely a French limited liability company i.e., société anonyme—with a share capital of 1000 million euros held by the Signatories in proportion to their investment shares (soon, many of them sold their shares and, in 2005, the major stakeholders established EUTELSAT Communications as the holding company [5: 308 and 309]).

On the other hand, EUTELSAT IGO was then created to assume the supervision of Eutelsat S.A. and became responsible for safeguarding the initial EUTELSAT

[5] "Exactly why EUTELSAT privatized is not as well documented as the other instances. Privatization of telecommunications services in Europe was occurring and the market was passing to commercial providers" [5: 308].

members' interests; such as to ensure that the company offers audiovisual services in conformity with the relevant national regulations and international agreements, on a basis of non-discrimination and fair competition [13]. For this purpose, EUTELSAT IGO may now—*inter alia*—participate, *via* its Executive Secretary who is elected by the Assembly of the Parties, in all meetings of the Board of Directors of EUTELSAT Communications (the holding company, listed on the Paris stock exchange) and Eutelsat S.A. (the operating company) [20]. In this context, the Greek State, which is an active member of EUTELSAT IGO since its accession in August 1987 [21], contributes actively to the work of all its organs.

3.1.1.2 Accession to EUMETSAT: Greek Law n° 1770/1988

In 1972, member States of the ESRO decided to implement the pre-operational programme METEOSAT,[6] with the aim to develop and use satellite technology for meteorological observation purposes (the decision also paved the way for the creation of ESA which "*inter alia* in the area of meteorology would become the organization to develop preoperational application satellites, as well as undertake operational activities in the same applications sector" [1: 191]). The first METEOSAT satellite was launched in November 1977 [22]; following on from that, in 1978, heads of national meteorological services decided to establish a European meteorological entity (and to agree on the relationship between such an entity and ESA) [1: 192].

The EUMETSAT Convention was agreed upon in Geneva, in May 1983, following completion of ESA's METEOSAT experimental program [23]. Hence, the aim was now to provide the European meteorological organizations with a framework for cooperation that would enable them to make the most out of space technology and at the same time to explore new space applications; in fact, "on the polar-orbiting front, the United States still operated a series of satellites but had indicated that from 1995, it expected others to share the burden of providing global coverage" [24: 54]. Thus, EUMETSAT, an international organization of sixteen States, including Greece[7] [24: 70 and 258], became a legal entity in June 1986 after adoption of its convention [25: 3]. The Convention establishing the *European Organization for the exploitation of Meteorological Satellites* (EUMETSAT) was

[6]The Meteosat satellite series were the first meteorological and geostationary satellites in Europe, which introduced the concept of a global system of geostationary platforms capable of observing the atmospheric circulation and weather around the equator, in near real-time. The Programme was approved by the ESRO in 1972. ESA was responsible for developing and operating the system on behalf of the *European Meteorological Satellite Organization* (EUMETSAT), which took direct operational control on 1 December 1995 as the *Meteosat Transition Programme* began [22].

[7]More precisely, the 16 founding States were: Belgium, Denmark, Finland, France, Germany, Greece, Italy, Ireland, the Netherlands, Norway, Portugal, Spain, Sweden, Switzerland, Turkey and United Kingdom.

ratified by means of the law n° 1770/1988, on April 1988 [26], and became part of national legislation in line with Art. 28 of the Greek Constitution.

Understandably, the Greek State additionally ratified the *Protocol on the Privileges and Immunities of the European Organization for the Exploitation of Meteorological Satellites (EUMETSAT)*, signed in Geneva on 1 December 1986 and entered into force on 5 January 1989, by means of the law n° 3019/2002 [27]. The protocol was adopted on the basis of Art. 13 of the EUMETSAT Convention, which laid down that EUMETSAT shall enjoy the privileges and immunities necessary for the performance of its official activity, in conformity with a protocol to be drawn up subsequently (*nb*. the convention was last amended in 2000 [25], and the protocol in 2004 [28]).

Regarding its purpose, EUMETSAT was founded as a dedicated organization aimed at establishing, maintaining and exploiting European systems of operational meteorological satellites, taking into account as far as possible the recommendations of the *World Meteorological Organization* [25: Art. 2 para 1]. This project was the first of its kind, given that, at that time, no international organization had arranged to provide Europe with meteorological data necessary to cover its zones of interest [25: Preamble]. As the safety of populations and the efficient execution of a wide range of human activities were, at that time and still are, dependent on meteorological data [25: Preamble], the need for accurate and reliable forecasts, based on observation data was already undeniable.

To this end, it was laid down that the EUMETSAT would take maximum advantage of the technologies developed in Europe, mainly in the field of meteorological satellites, by providing for operational continuation of the programs that had proved technically successful and cost-effective [25: Art. 2 para. 3.a]. The new entity would also rely as appropriate on the capabilities of the existing international organizations carrying out activities in a similar field [25: Art. 2 para. 3.b] and contribute to the development of space meteorology techniques and meteorological observing systems using satellites [25: Art. 2 para. 3.b], with the ultimate goal to improve services at low cost.

In this context, the EUMETSAT Convention, Art. 4, mentioned that the Council—namely, the EUMETSAT supreme decision-making body ("The Council shall have the powers to adopt all the measures necessary for the implementation of this convention" [25: Art. 5 para. 1])—may be composed of no more than two representatives for each member State, one of whom must be a delegate of his country's meteorological service; however, the representatives may also be assisted by advisers. As a result, Presidential Decree 161/20.06.1997, Art. 5 (ζ), laid down that the Commanding Officer of the Greek Meteorological Service to the Ministry of National Defense [*Διοικητής της Εθνικής Μετεωρολογικής Υπηρεσίας (ΕΜΥ) του Υπουργείου Εθνικής Άμυνας*] shall be the responsible authority for setting up collaborations between the Greek State and any relevant international organization, such as amongst others EUMETSAT, and for acting as a legal representative for the country [29]. According to the same Presidential Decree—i.e., Presidential Decree 161/20.06.1997, Art. 14.2 (στ)–, the Meteorological Satellites RADARS Service [*Τμήμα 6 Μετεωρολογικών Δορυφόρων RADARS*] is the specialized agency responsible for specifically handling issues related to the use and exploitation of EUMETSAT satellites by the Greek State. It is interesting to note that, for transparency reasons, all the names of the delegates and advisors to the EUMETSAT Council are published in the EUMETSAT Annual Report; therefore, the Greek government's delegates and advisors at the EUMETSAT are likewise listed there [30, 31].

The administrative and operational expenditure of EUMETSAT is to be covered by the financial contributions of member States as well as by any other EUMETSAT income. In this context, it was also clarified that "The Council may accept voluntary contributions, whether in cash or otherwise, provided they are made for purposes compatible with the objectives, activities and principles of conduct of EUMETSAT" [25: Art. 10]); that for MOPs (Meteosat Operational Programmes) as these are defined in Annex I of the Convention, each member State would pay an annual contribution in convertible currency [25: Art. 10 para. 6], etc. In more practical terms, the contributions of each State, including Greece, are detailed in the EUMETSAT Annual Report [31: 90].

It is beyond doubt that the Greek State benefited considerably from its participation in all the above-mentioned structures, namely the EUMETSAT and the EUTELSAT, and was able to provide its citizens (as well as its governmental and non-governmental entities) with high quality space-related services. However, soon, it also began to gradually establish and improve its presence in the context of the core European Space Agency (ESA) framework, which is pursuing a much wider objective (especially in comparison to the ELDO and the ESRO which had more narrowly defined purposes [1: 161]).

3.1.2 Joining the ESA Framework and Acquiring Full ESA Membership

The European Space Agency (ESA) resulted from the merging of the ELDO [4] and ESRO [32], in 1975, and had from the outset a crucial role in defending the scientific and technological interests of the European space actors, while contributing to the objective of ensuring the continuity of European space activity. Today, space activity has entered a new phase (i.e., "the focus is shifting from space systems to space applications and there is now strong political will to further develop European downstream industry" [7: 8]) thereby highlighting ESA's enduring role as *Europe's gateway to space* [33].

As regards Greece, one may note that—*stricto sensu*—cooperation with the ESA began in the early 1990s, when the Greek State signed its first cooperation agreement with the Agency [34]. Literally, it agreed upon a first protocol of cooperation with the ESA on 4 July 1994, paving the way for the regular exchange of information, the award of fellowships, joint symposia, mutual access to databases and laboratories, as well as studies on joint projects in fields of mutual interest (the Agreement was concluded for a period of five years with the possibility to be renewed [35]). Since then, the ESA established a number of fruitful collaborations with Greece, and additionally agreed to provide the Greek Ministry of Transports and Communications with consultancy support to the Hellas-Sat project [35].

Against this background, the following paragraphs aim at giving an overview of the steps taken by the Greek State to strengthen its involvement in the framework of ESA.

3.1.2.1 Principal Measures Taken Towards Becoming a Full Member of ESA

The ESA was created in 1975, as an inter-governmental organization; in reality, the ESA convention was signed on 30 May 1975, but it came into force "only after a period of more than five years when, on October 1980, in accordance with the provisions of the ESA convention, with the deposition of France's instrument of ratification the required number of countries had completed the formalities for ratification or adherence" [1: 160]. Hence, its purpose is to provide for, as well as to promote for exclusively peaceful purposes "cooperation among European states in space research and technology, and their space applications [2: Art. 2]". To achieve this goal, ESA has a range of instruments at its disposal. In particular, it may *inter alia* reach agreements aimed at gradually integrating new member States into the ESA framework, and undertake a series of actions and initiatives considered to be necessary and appropriate for the fulfillment of its objectives.

Interestingly, Greece joined ESA almost a decade after joining the main European management (or organizational) structures for space services, namely EUMETSAT and EUTELSAT presented above. In essence, it is all the resolute steps taken by the Greek State in this direction which effectively allowed it to finally develop a domestic space sector.

3.1.2.1.1 First Collaboration Agreement with the ESA in the Early 90s

As already noted, the first Protocol on cooperation between the Greek State and the ESA was agreed upon in 1994 (it allowed it to participate in specific ESA activities and programmes), on the basis of Art. XIV of the ESA convention [2] mentioning that "The Agency may, upon decisions of the Council taken by unanimous votes of all Member States, cooperate with other (...) Governments (...) and conclude agreements with them to this effect". However, literally, said cooperation was given its real impetus only after adoption of the cooperation agreement signed on 17 January 2001, ratified by means of the law n° 2953/2001 [36] and entered into force on 13 November 2001 [as mentioned in 37]. Importantly, right from the outset, the focus was put on the objective to bring about a gradual development of space activities in Greece and to also foster cooperation between the parties [36: Art. 1]. Parallel to that, the law established the competence of the General Secretariat for

Research and Technology [*Γενική Γραμματεία έρευνας και Τεχνολογίας*] in this field, under Art. 2. which was further mentioning that a committee would be set up—by a decision of the Minister for Development [*Υπουργός Ανάπτυξης*]—to be responsible for the coordination and monitoring of the implementation of said agreement [36: Art. 2].[8]

In this context, it was stipulated that parties would keep each other informed "on all their respective activities and programmes" (i.e., Art. V para. 3 of the ESA Convention, titled 'Activities and Programmes', mentions that "With respect to the coordination and integration of programmes referred to in Article II c, the Agency shall receive in good time from Member States information on projects relating to new space programmes"), "and on their progress and shall consult regularly (. . .) on areas offering potential for cooperation and participation by Greece in an Agency programme" [36: Art. 2]. Furthermore, the agreement made particular reference to specific areas that were considered by the parties as offering potential for cooperation [36: Art. 2.2][9] while it was also established that parties would exchange information on specific topics, such as the content and timing of their present and future space programmes, as well as regarding matters of scientific and technical interest that may arise out of their space activities [36: Art. 2.3]; such clause was based on Art. III of the ESA convention, laying down that ESA Member States undertake to facilitate the exchange of scientific and technological information, provided such thing is necessary and related to space activities and space technologies which are under the responsibility of ESA [2: Art. III].[10]

Finally, it was established that the agreement would remain in force for a period of five years [36: Art. 9.3]. At the same time, it was concluded that the parties would consider, one year at least before the expiry of said agreement, how such cooperation could be further developed and "in particular examine the possibility of Greece being granted the status of Associate Member or Member State of the Agency" [36: Art. 10].

[8]Furthermore, it was established that the General Secretariat for Research and Technology [*Γενική Γραμματεία έρευνας και Τεχνολογίας*] would undertake to pay the annual contribution of the Greek State for participating in ESA programmes [36: Art. 2].

[9]In particular, mention was made to space sciences (astronomy, astrophysics, solar system exploration and solar-terrestrial physics); Earth Observation research and application (environmental monitoring, meteorology, aeronomy and geodesy); telecommunications (service demonstrations and satellite navigation); microgravity research (space biology and medicine, and materials processing) and Ground Segment engineering and utilization.

[10]However, it was underlined that a Member State "shall not be required to communicate any information obtained outside the Agency if it considers that such communication would be inconsistent with the interests of its own security or its own agreements with third parties, or the conditions under which such information has been obtained" [2: Art. III].

3.1.2.1.2 The 2004 Agreement of Accession—Entered into Force in 2005

Based on Art. XIV para. 1 of the ESA convention mentioned above and on the evolutionary pattern that is followed by the Agency, Greece joined the ESA—in 2005—first "under a protective regime, participating in specific projects with local entities, so that [*it could*] adjust to the operation of the agency" [38]. Interestingly, the Greek State skipped the interim status of Associate Membership and moved from cooperation agreements directly to full ESA membership [39].

More precisely, before completion of the five years period which had been agreed in the 2001 agreement,[11] the Greek State formally applied to join the ESA [40] (that is, in September 2003). Following negotiations, an agreement on the accession of Greece to the ESA was signed on 19 July 2004, and later ratified by the Greek Parliament by means of Law n° 3308/2005 [37]. Said accession became effective on 9 March 2005 [41]; hence, Greece acquired the status of the 16th ESA Member State [42] on the basis of Article XXII of the ESA Convention [2], stipulating that "After the entry into force of this Convention, any State may accede thereto following a decision of the Council taken by a unanimous vote of all Member States". At the same time, the Greek State adopted the conditions governing the functioning of ESA [37: Art. 2.2] and became party to the Security Agreement on the protection and exchange of classified information.[12]

Needless to say that the 2004 agreement of accession precisely laid down that from the date of its entry into force for Greece, the 2001 cooperation agreement would be considered to be null and without legal effect [37: Art. 5]. This would in particular apply to agreements or understandings reached between Greece and the ESA, in reference to its participation in ESA optional programmes [37: Art. 5].

3.1.2.1.3 Participating in the ESA Procedures and Governing Bodies

As part of its objective to shape the development of Europe's space capability and infrastructures, as well as to ensure that space investments continue to deliver benefits to EU citizens, ESA coordinates the efforts of its member States and undertakes space programs and activities far beyond the capacity of each country and space organization. In this context, it aims at promoting inter-State collaborations (ESA acts in this case as a mechanism of international cooperation between its

[11] The five years period agreed upon in the 2001 Agreement of Cooperation began after notification by the Greek State of the approval mentioned therein [36: Art. 9.1], namely on 13 November 2001—as outlined in Law n° 3308/28.01.2005 entitled "Ratification of the agreement between the Government of the Hellenic Republic and the European Space Agency etc." [see 37]—, roughly covering the period 2002–2006.

[12] According to Art. 1 of the Security Agreement, 'classified information' is information of any description, documents or materials, which in case of unauthorized disclosure could harm the interests of one or more contracting parties and are designated as such on the basis of the security classification [43].

Member States) while it may also create partnerships with other entities, as an actor of international cooperation [44].

Such initiatives are decided and implemented by the ESA governing bodies, namely (i) the Council of Member States -or ESA Council- and (ii) the Director-General. The *ESA Council* is the body which is representing member States; it is composed of either the Member States competent ministers or their delegates (each delegation may include a maximum of two persons),[13] and works on the basis of a one-country one-vote principle[14] regardless of the State's size or financial contribution [45]. Correspondingly, the *ESA Director-General* is its legal representative, who is empowered to represent it in all its acts, as assisted by his staff.

Hence, by becoming a full ESA member, the Greece State first established, and then strengthened its involvement in the governing bodies mentioned above. Indeed, it was initially laid down (i.e., in the 2001 cooperation agreement) that "(f)or items concerning relations between the Agency and the European Union, Greece may upon invitation from the Agency take part as an observer in meetings of the council and of its subordinate bodies" [36: Art. 2.6]. Furthermore, it was stipulated that "the parties shall consult on matters of common interest appearing in the agenda of meetings of international bodies" [36: Art. 2.7]. Nonetheless, as it was reasonable to expect, the agreement on the accession of Greece to the ESA Convention adopted in 2004 went one step further, granting that "As from the date of accession (i.e., of the Greek State to the ESA convention), the provisions of the convention including all the measures adopted by the council, shall be binding upon Greece and will apply in this Member State" [37: Art. 3.2]. Additionally, it was highlighted that the Greek State will be treated in the same way as other member States with regard to decisions, regulations, resolutions as well as any other acts that would be carried out by the council [37: Art. 3.2].

Finally, the 2004 agreement of accession laid down in detail the financial participation of the Greek State [37: Art. 7]. Such arrangement was adopted taking into account that ESA's proper functioning is ensured by the stability of its budget (consisting of obligatory contributions from the ESA member States[15]), in addition to a long-term planning of space activity—combining missions of a different scope as well as ensuring balance between the disciplines—and a complementarity of responsibility between the ESA developing the platforms, and national laboratories developing on-board instruments [46].

---

[13] As an international organization, ESA's council is formed by the representatives of the Member States. However, the council may also meet at ministerial level, a political role formerly executed by the European Space Conference [1: 161].

[14] According to the ESA Convention, Art. XI (titled 'The Council'), each Member State is empowered to decide on new and continuing programs; however, a State has the right to vote -and a say- only on issues that affect accepted programmes in which it takes part [2: Art. XI para. 6 a].

[15] ESA consists today of 25 Member States (in addition to Canada), while "each country contributes to its budget according to its GDP, with the most important funding coming from the founding members", [45].

3.1.2.2 Acquiring Knowledge by Leveraging ESA Programmes and Initiatives

The ESA is empowered to adopt a range of measures to attain its objectives. Hence, to play a role in space policy formulation,[16] it may promote specific initiatives in the field of space, through the elaboration and implementation of activities and programmes [2: Art. II (b)], as well as an appropriate industrial policy; at the same time, it may propose its Member States to adopt all the measures deemed necessary to realize its targets [2: Art. II (d)]. On this basis, ESA activities are divided into mandatory—that is, activities in which all member States participate—and optional ones, as further defined in the ESA Convention [2: Art. V.1]. Furthermore, the ESA may carry out operational activities in the area of space applications, under conditions which are defined by the ESA Council, by a majority of all member States [2: Art. V.2].

Regarding the involvement of the Greek State in ESA space activities, it should be noted that the 2001 Agreement of Cooperation made then specific reference to "areas with potential for cooperation between ESA and Greece *including* space science, Earth observation research and applications, telecommunications, satellite navigation, microgravity research, and ground segment engineering and utilization" [47]. As a result, at the time of its accession to the full ESA membership, Greece was already actively participating in ESA's telecommunication and technology activities, and in the Global Monitoring for Environment and Security Initiative [42]. More precisely, said agreement enabled the country to participate in the ARTES telecommunications program, the GMES (widely known as Copernicus) and the GSTP (i.e., Earth observation program and General Support Technology Program), as well as to familiarize with ESA's operational procedures [48, 49].

The 2004 agreement of accession allowed for a more pronounced involvement of the Greek State in ESA programmes, while all the previous agreements establishing its participation in ESA optional programmes ceased to apply [37: Art. 5 already referred to]. Literally, the two parties agreed, by signing the new agreement of accession, that the rights and obligations of Greece as a State being involved "in the programmes above"[17] would be governed by the respective arrangements and by all the applicable rules [37: Art. 5].

Participating in the ESA programs provided the country with tangible benefits and knowledge, as well as valuable opportunities. By way of illustration, the national space sector strongly benefited from specific actions, like ESA BIC Greece [50][18] or

[16] The ESA is empowered to elaborate and implement a long-term European space policy, by recommending space objectives to the Member States, and by concerting the policies of the Member States with respect to other national and international organizations and institutions [2, Art. II para. (a)]. In addition to that, the ESA is entitled to coordinate the European Space Program and national space programs, and integrate the latter progressively into the first [2: Art. II para. (c)].

[17] Specific reference was made to ESA optional activities (according to the ESA Convention, Art. V, *all* Member States participate in the mandatory activities).

[18] ESA BIC Greece was established with the purpose to create and strengthen the community of space-related start-ups.

the CubeSat projects (i.e., whereby seven CubeSat missions "that demonstrate a variety of services including connectivity and secure communications are being developed by small and medium-sized companies and universities in Greece, following an open call and selection led by the ESA on behalf of the Greek Ministry of Digital Governance" [51]). In this context, it is undoubtable that the Greek space sector progressed to a higher level in its development phase, largely due to the opportunities (such as mainly knowledge and experience gained) unlocked by participating in the ESA [52; and as a point of comparison 53]. Taking this into account, the Greek State evidently sought to continue strengthening its involvement in ESA activities and programmes—and continues to do so [54].

3.2 Greek Space Activity Seen in the Light of Developments in the EU Space Policy

The ESA and the EU are the two principal international European bodies [55] in reference to space matters. In truth, although a close analysis of the two systems reveals that they are far from sharing the same approach to space collaboration—mention was made, for example, to the "*clash* between the ESA principle of fair return and the EU premise of free competition"[19]–, the two entities should be considered as highly complementary. Therefore, to the information above should be added a presentation of developments taking place in the framework of the EU, as said developments also have an impact on member States space policies (which includes Greece).

Before examining in more detail some of the major instruments of the EU space policy, it is necessary to note that the Greek State's alignment with said policy is founded on the country's EU membership: Greece is a full member of the European Union (EU)—former European Community (EC)—created in 1957 following adoption of the Treaty of Rome (i.e., *Treaty establishing the European Community*), on the basis of its instrument of accession to the former EC signed in May 1979, in Athens, and ratified by the Greek Parliament a few days later, on 28 June 1979. Thus, since its accession which became effective on 1 January 1981 [56], Greece is bound by *all* the EU decisions and legislation, in line with national and European law. In this context, the Greek State soon became actively involved in the EU effort to develop a competitive space legislation and policy, and to strengthen collaboration between member States on space matters. Parallel to that, it took careful account of the EU priorities in space, while developing its own legislation and space policy, building further on the (practical and scientific) knowledge that it had gained

[19] "It cannot be denied that the European space industry would not be as competitive and efficient as it is today if it were not for the fair return principle—this has allowed many European states to gain expertise and know-how in the field, which otherwise would have remained occupied by a few players in traditional European space powers like France, Germany, Italy and the UK" [45].

thanks to its involvement in several ESA space programs. Such experience and participation allowed it to finally put the emphasis in developing its own space capabilities.

3.2.1 Putting the Focus First on Developing European Space Capacities

In essence, the EU was empowered to adopt a common policy on space matters—and to commit member States to take, within their domestic legal systems, the measures necessary to align with EU priorities—only following enactment of the Lisbon Treaty, adopted in 2007 and entered into force in 2009. Yet, literally, the development and implementation of a common EU approach in the field of space activity and exploitation mainly started in the early 2000s; which also inspired and encouraged Greece to make its first steps in this domain, as briefly outlined below.

3.2.1.1 First EU Instruments and First Steps for Greece in Space Activity (Early 2000s)

European States took note of the importance of space and space-related activity more or less around the 70s [57][20] and 80s [58]. Following on from several initiatives and studies [e.g., referred to in 59], said awareness resulted in the adoption of a *first* set of key EU policy instruments, starting from 2003, such as the *White Paper on European Space Policy* [60]. In fact, the *White Paper* was building further on the results of a *Green Paper* titled "European Space Policy" which had marked the beginning of an official consultation period on options for European activities in space [61], having established that "no European nation is capable of independently maintaining a space policy at the necessary level" [61: 4]. These *first* EU policy instruments essentially highlighted that "the nature of modern technology made it essential to place space policy issues at the center of the EU's efforts to integrate Europe" [58: 45]; although, in truth, a number of (military) satellites had been launched and were operated by European States since the late 90s, in a satisfactory way in the context of their own policies [62].

More precisely, the release of the *Green Paper on European Space Policy* dated 21 January 2003 [61] was motivated by a desire to achieve, then, EU autonomy on space matters, in a manner that would be demand driven and in support of other EU

[20] "Already in 1979, the European Parliament (EP) adopted a "Proposition for a Resolution on the Community's participation in space research" and later, in 1981 the EP adopted a resolution on a "European Space Policy" for the first time. The first reference to national space legislation appeared in 1987 when the EP urged the European Commission to contribute to the process of development and codification of space law in order to avoid a time-consuming codification later on" [57: 36].

initiatives. The goal was to additionally enhance EU's role as a key space actor, based on existing capacities (ex. Galileo and the GMES initiative) considered useful for the conduct of other common policies as well; such as for security purposes, or in the context of the Common Foreign and Security Policy (CFSP) and the European Security and Defense Policy (ESDP) [61: 4]. Following on from that, the *White Paper on European Space Policy*, adopted on 11 November 2003 [60], was a call for action to all stakeholders: like, namely, the European Space Agency (ESA), EU and ESA Member States with their national space agencies, research centres, and the industry. As a result, said instrument established that the European Space Policy would be implemented within the context of a multi-annual European Space Programme (ESP) as a strategic agenda for Europe, and would create a mechanism for achieving such targets [60].[21] Overall, the principal purpose was to ensure (and maintain) an independent access to space resources, to the benefit of the EU and on an affordable basis for all member States.[22]

Hence, the building of a new form of governance for space activity was based on a two-phased approach. The first phase (2004–2007) consisted of implementing the topics covered by the Framework Agreement that was at that time signed between the EU and ESA (in 2003), so as to enable them set common objectives and mount joint initiatives. More decisively, the second one (2007–onwards), programmed to begin with the coming into force of the proposed European Constitutional Treaty, was aimed at establishing space as a shared competence between the EU and Member States.[23]

Interestingly, the period above, when the EU started to actively address space policy matters is also the one when the Greek space sector and activity began to really emerge. Indeed, *Hellas Sat SA*. was founded in 2001—in Greece—as a subsidiary company of "Hellas Sat Consortium Limited" that had been established in Cyprus (i.e., having its central administration, registered office and principal place of business there as well) with the specific objective of exploiting the "satellite geostatic orbit of 39 degrees east, and other relative frequencies that would place Greece among countries with direct access to a satellite and strengthen OTE's position as a telecommunications operator" [63]. Hence, said company was given the exclusive right to utilize the above orbital slot that was granted to Greece and Cyprus (namely, 39 degrees east as well as its associated radiofrequencies), in order

[21]The mechanism would allow determining priorities and identifying existing technological and financial gaps, setting objectives and priority activities, allocating roles, tasks and responsibilities of the different actors and framing annual budgets [60: 4].

[22]The White Paper also gave an account of existing space activities—and of applications under development—that could be used to support EU objectives and policies; in addition to that, it recommended space capacity-building actions and outlined some necessary changes in the governance and financing of EU space activities [60: 24].

[23]In this context, the White Paper gave a clear roadmap for the recommended actions in all the above-mentioned domains, such as in support of projects within the enlarged EU (e.g., Digital divide; GMES etc.), to allow extending and strengthening space policy (e.g., space technology; space exploration etc.) also in relation to governance and resources [60: 36 et seq.].

to provide "telecom services such as, Digital T.V., fast Internet, VSAT services, Telephone Traffic and in general interactive services [*and cover*] the biggest part of 2004 Olympic Games TV Broadcasting needs" [63], through the leasing (then) of the DFS Kopernikus 3 satellite from Deutsche Telekom used under the name *Hellas Sat 1* [64]. Hellas Sat 1 was exploited by the Greek State until 2003, when the latter finally launched its (own) first satellite, *Hellas Sat 2*.

More precisely, the first Greek and Cypriot Telecommunications Satellite[24] named *Hellas Sat 2* (or *HELLAS-SAT II*) with an expected life of 15 years, was launched from Cape Canaveral on 13 May 2003, after several years of preparatory work[25]; in this manner, Greece essentially confirmed that it attached the highest priority to also developing its own space technology [49: 51]. Overall, in the first years of its existence, *Hellas Sat SA* "managed to procure, launch and operate one of Europe's most powerful communications satellites with its two ground stations in Greece and Cyprus" [65]; which allowed experts to argue that "(t)he gains are not only financial, but also prove that even the smallest of countries can join the small and elite group of spacecraft owners and satellite operators" [65]. In other words, the first efforts made in this area, as well as the results that were obtained, were quite promising.

3.2.1.2 After Adoption of the Lisbon Treaty & First Space Council Works (Until 2010)

As the pace of changes accelerated sharply until the end of 2010, the need appeared for the EU to have a greater participation in space matters by developing a more active role. On this basis, it focused its effort in two specific directions. *On the one hand*, it aimed at developing legal instruments that would facilitate the exploitation of space systems and resources to the benefit of EU policies –such as, the promotion and/or adoption of legislative, regulatory and standardization measures in this field–, coupled with the funding and carrying out of new initiatives related to the exploitation of space and the strengthening of public-private partnerships. *On the other hand*, it went into establishing a framework for efficient cooperation with the ESA, whereby said entity would act as an implementing agency for the development and procurement of the space segment and ground segment involved in several EU initiatives [66]. As it was reasonable to expect, said developments encouraged further the Greek State to also strengthen its own involvement in space matters.

---

[24] "Following the allocation of frequency bands and a geosynchronous slot by the ITU, the newly formed Cyprus-based HELLAS SAT Consortium LTD and its subsidiary Greece-based HELLAS SAT S.A., initiated talks with EADS-ASTRIUM for the manufacturing of a Eurostar 2000+ telecommunications satellite. The contract with the manufacturer was signed on July 11, 2002" [65].

[25] "The process for deploying a telecommunications satellite began with the Governments of Greece and Cyprus initiating coordination with the International Telecommunication Union (ITU) in 1993 and 1995 respectively for assuring frequency bands at the 39o E geosynchronous slot" [65].

3.2.1.2.1 The Major Developments in EU Space Policy

The EU was essentially empowered to coordinate Member States activities in outer space only following adoption of the Lisbon Treaty [67], which entered into force in 2009 and established space as a shared competence between the EU and Member parties. Literally, the *Treaty on the Functioning of the European Union* (TFEU) stipulated, in Article 189.1—which was in fact reiterating Art. III-155.1 of the draft Constitutional Treaty [68]—, that "(t)o promote scientific and technical progress, industrial competitiveness and the implementation of its policies, the Union *shall draw up a European space policy*. To this end, it may promote joint initiatives, support research and technological development and coordinate the efforts needed for the exploration and exploitation of space" (emphasis added). Furthermore, Art. 189.2 clarified that to reach this objective, "the European Parliament and the Council (. . .) shall establish the necessary measures, which may take the form of a *European space programme*" (emphasis added).

However, although the EU was then endowed with the necessary powers to perform its regulatory functions in an efficient way and to the benefit of European space actors (i.e, EU institutions were thus "able to assert their jurisdiction *in re* competition in *any* field of space activity" [69]) it faced, in reality, difficulties in asserting competition jurisdiction in this particular context [69]. As an explanation, it is—on the one hand—contended that the "classical market approach of the EU i.e., through harmonization, may not be the best suited to space activities" [57: 40]; while, on the other hand, it is argued that the EU competence in space matters has been "curtailed by the exclusion of any harmonization in space which leads decision makers to the search for alternative means for the creation of the much-praised coherent approach to space law in Europe" [57: 5]. As a result, the focus was soon put on establishing first an improved regulatory dialogue, as well as convergence on key space matters.

In this context, the EU—which was until then principally concerned with the improvement of space systems[26]—prioritized strengthening its normative role, and proposed *inter alia* the adoption of a regulatory framework for space activities [70]. For instance, it released a draft *International Code of Conduct for Outer Space Activities (ICoC)*[27] in 2008 [71], to mainly promote transparency at the operational level[28] (e.g., such as by enhancing "the safety, security, and sustainability of all outer space activities pertaining to space objects, as well as the space environment" [73]), but also to eventually clarify how sustainability may be applied in space, where it seems that the nature of acceptable activities is not always

[26] Such as the sustained investment in developing the three space programmes: i.e., the European global navigation satellite systems—European Geostationary Navigation Overlay Service (EGNOS) and Galileo—and the earth observation programme Copernicus [70], see.

[27] "This initiative, led by the EEAS, was launched in 2007 in response to the UN Secretary-General's call for concrete TCBMs for space" [70].

[28] For instance, "(. . .), the Code of Conduct does not overlap with, but is complementary to the 1975 Registration Convention" [72].

apparent.[29] As a few revisions of the draft were carried out considering the comments received from several other States [72], so as to encourage the exchange of views at international level, the EU did establish itself as one of the primary actors in the international space debate [72].

3.2.1.2.2 The Works and Impact of the European Space Council

During the same period, the two entities competent for European space matters (i.e., the ESA and the EU) joined forces. In particular, a cooperation was established between them following adoption *first* of the EU resolution on a European Space Strategy, on 16 November 2000 (whereby the EU Council requested the EU Commission "to set up, as soon as possible, in cooperation with the ESA and by the end of 2000 at the latest, a joint high-level Task Force involving the Commission and the ESA Executive" [66]) and, a few days later, of a *framework agreement*. The latter was signed between the ESA Council and the EU Council, on 25 November 2003 [74], and it aimed at reaching "cooperation between the Parties, while maintaining their respective distinct tasks and responsibilities (Art. 2)" and at developing a coherent overall European Space Policy. In this context, Art. 8 of the framework agreement also created a coordinating body—i.e., the *European Space Council*–, to provide orientations supporting the achievement of the agreement's objectives, make recommendations and advise the parties.[30]

Literally, "(t)he approval of the Space Council meant that the new space policy now represent(ed) official policy for all the member states of the ESA and EU" [58: 42]. Thus, the *European Space Council* (which met for the first time on 25 November 2004) offered ministers representing the EU and ESA Member States the opportunity to discuss the adoption of a European Space program, in the footprint of the treaty establishing a European constitution; its aim was to make it clear that space facilitates everyday life in many aspects, and that the space sector is one of the fastest growing economic sectors [75]. At the same time, key challenges were examined, such as the strategic importance of the flagship initiative for Global Monitoring for Environment and Security (GMES) [76]; or next initiatives that had to be considered [77, 78], including the need for enhanced cooperation between space and non-space sectors, and between the space industry, leading research organizations and universities, to reap the high innovation potential of space research and innovation; in addition to the sustainable use of GALILEO and GMES systems [79]. Even More importantly, the *first EU space policy* was formalized in May 2007 and adopted

[29] Detailed rules are laid down, such as that States should minimize the risk of accidents in space, or collisions between space objects (Art. 4.1); or refrain from any action which brings about, directly or indirectly, damage or destruction of space objects, unless such action is justified (Art. 4.2) [73].

[30] Precisely, the ESA—EU Framework Agreement mentions, in Art. 8: "The coordination and facilitation of cooperative activities under this Agreement shall be accomplished by regular joint and concomitant meetings of the Council of the European Union and of the Council of ESA at ministerial level (Space Council)".

during the 4th European Space Council [80], in which ESA and the EU decided to improve coordination of their joint action and programmes, and to also clearly define their respective roles and responsibilities on space matters [81].

However, it appears that discussions and working sessions within the council were interrupted after some time. The *European Space Council* met again years later, in 2019, and put particular emphasis on encouraging the ESA and the EU to implement their respective space activities and programmes. The main challenge was now to strengthen Europe's role as a global actor, while promoting international cooperation in this field [82], as well as trough contributing to the development of a common approach to the space matters, which later resulted in the adoption of the *Matosinhos Manifesto* on the acceleration of the use of space in Europe [83]. In essence, the works of the European Space Council and mainly the decisions it took contributed further to raising awareness on the key importance of space activities for both the EU and its member States, including Greece.

3.2.1.2.3 Initiatives Adopted by the Greek State During this Period

For the Greek State, this period was characterized by its own growing involvement in ESA space programmes (i.e., as highlighted above, Greece acquired full membership in 2005). However, it had not yet established a specific legislation to regulate space activities conducted under its jurisdiction (interestingly, it was noted, in 2009, that "so far only a handful of countries have adopted space laws as such. Among them there are only five EU Member States (. . .)" [57]). In this context, it succeeded to launched its first telecommunications satellite owned together with Cyprus and did fulfill all of its obligations under international law; such as was to inform the Secretary-General of the United Nations on the launching of its space object (i.e., in accordance with Art. IV of the Convention on Registration of Objects Launched into Outer Space, it had communicated all the information relating to the launch and operation of *Hellas Sat 2* [84]).

The period under review is further characterized by other important decisions, such as the adoption of a cooperation agreement—on 31 August 2006, in Athens *inter alia*–, signed by the Greek government with France, Belgium, Italy and Spain (all represented by their ministers of defense). Said agreement established the Greek participation in the development of *Helios 2* i.e., an Earth Observation system being used for military applications developed by France together with Belgium, Spain and Denmark [85] (*Helios-2* was the successor to the *Helios-1* military Earth Observation program, developed by France in collaboration with Spain and Italy). The agreement, ratified by means of law n° 3546/2007 [86], laid down that participants agreed to jointly examine any technical or operational development questions that would result in the improvement of the *Helios-2* programme [86: Art. 1.2]. The content of works and studies to be carried out under that agreement was described in detail, including those intended for the benefit of the Greek State [86: Art. 2]; in return, the latter obtained access to the *Helios-2* system as it was precisely laid down in said agreement [86: Art. 6].

Until roughly the end of 2010, Greece had undoubtedly made all the necessary efforts to enable its domestic space sector to take off, taking into account all relevant national and industrial specificities. As a matter of fact, it had strongly (and quite efficiently) promoted external collaborations and synergies, relying on the expectation of mutual expertise. Nevertheless, it is the next decade which was the real period of transition for the Greek State, as it then eventually established all the necessary legal and institutional frameworks that allowed it to essentially improve its position in a strongly competitive context.

3.2.2 *Towards a New Era for the Greek Space Sector: From 2010 to 2020*

The years 2010–2020 are viewed as a decade of profound change and measurable progress [87]: developments highlighted the importance of outer space for human activities even more, due to ground breaking advances in fields like space robotics and technology, as well as in reference to their applications. This resulted in an even greater participation of private actors, a further commercialization of space and the emergence of new space-faring countries. In this context, discussions focused *inter alia* on issues such as strengthening security in and from outer space, and the coordination of space efforts and space-related policies.

3.2.2.1 Principal Developments at the EU Level

First, on the legislative front, the European Commission released the *Space Strategy for Europe* in October 2016 [88], regarded to have two main dimensions: "policy for space, focused on the promotion of scientific and industrial competitiveness, and space for policy, addressing the impacts of the EU's space investment on the EU's political priorities" [89]. Hence, specific priority areas for action included maximization of the benefits of space for the EU society and the EU economy; the fostering of a globally competitive and innovative EU space sector; the reinforcement of Europe's autonomy in accessing and using space in a secure and safe environment; and the strengthening of EU's role as a global actor as well as in promoting international cooperation [89].

A few years later, the *EU space programme for 2021-2027* was adopted, based on the political agreement finally reached between the European Council and the European Parliament in December 2020. The final act was signed in April 2021 and entered into force on 12 May 2021, as Regulation (EU) 2021/696 also referred to as the *Space Program Regulation* [90]. Said instrument simplified the EU legal framework and governance system, which was now consisting of (i) the current flagship space programmes—i.e., Galileo, EGNOS and Copernicus— which were brought together under one single umbrella[31] and (ii) two new components: the Space and Situational Awareness (SSA) programme and the new Governmental Satellite Communication (GOVSATCOM) initiative (aimed at monitoring space hazards and providing national authorities with access to secure satellite communication) [89]; in this specific context, the Greek State acquired (in 2019) its operational governmental satellite communications system, which allowed the interconnection of Government and Security Users, referenced as GreeCom (Greek Governmental SatCom) [91].[32]

At the institutional level, it is noteworthy that the *European Union Agency for the Space Programme* (EUSPA) was created, building on the legacy of the European GNSS Agency, although with increased responsibilities.[33] In addition to that, the collaboration between the EU and ESA went one step further, as a *financial framework partnership agreement* was signed on 22 June 2021 (between the EU Commission and ESA). Said agreement created a new context for cooperation, both between these two bodies and with the new *European Union Agency for the Space Programme* (EUSPA) [92].

3.2.2.2 National Measures Adopted—During this Period—on Space Matters

As expected, the decisions and measures adopted at EU level had a decisive impact on national policy choices in the field of space, leading to the creation of legal and institutional frameworks necessary for the conduct of space activities. The measures adopted by Greece during this particular time frame are essentially presented and analyzed in Part II, which is dedicated to presenting the Greek Space Law, namely (Chap. 4) presenting the Competent State and non-State Greek authorities and (Chap. 5), introducing the Greek legislation on space activities. However, for the sake of completeness, it is opportune to mention, here as well, at least some of the major initiatives in the order they were adopted, and to briefly refer to a few key points of the *Greek space policy* as many of these developments were prompted by the steps taken at the EU level.

[31] With the purpose "to ensure the continuous provision of high-quality secure space-related data, information and services, to maximize the socio-economic impacts of the investments, and to enhance the security of the EU while also promoting its role in the international arena" [89].

[32] GreeCom is based on the exploitation of Greek frequencies and orbital position of the HellasSat satellites, as operated by Hellas Sat S.A. [91].

[33] "EUSPA is entrusted with the exploitation of EGNOS and Galileo, and with the overarching coordination of user-related aspects of GOVSATCOM. It is also expected to implement activities in support of the development of downstream applications stemming from the programme, activities that would ensure the security accreditation of all the programme's components, and other communication-related and market development activities" [89].

3.2.2.2.1 Enacting the Greek Space Legislation – Setting Up the Greek Space Authority

First and foremost, Greece adopted—in 2017—its national space legislation, namely, law n° 4508/2017 titled *Authorization of space activities—Registration in the National Registry of Space Objects—Establishment of the Hellenic Space Agency and other provisions* [93]. More precisely, the substantive provisions of the law encompassed rules necessary to regulate *inter alia* topics such as the scope of its application; issues related to the development, setting up and implementation of a space activity (including its authorization, supervision etc.) as well as risks and hazards that may occur during the operation of objects in space; they are all presented, in detail, in Part II of this study.

In addition to that, law n° 4508/2017 established the national space agency (under the name, then, of Hellenic Space Agency S.A.), which was later reshaped into *Hellenic Space Center* following the adoption of law n° 4623/2019. Hence, the principal operating rules of the national space agency were laid down in 2017, allowing the new authority to play a significant role in addressing space matters, together with a few other State and non-State (space and space-related) authorities.

3.2.2.2.2 Launching Additional Greek Satellites: *UPSsat*, *Hellas Sat 3* and *Hellas Sat 4*

In the context of efforts undertaken by stakeholders to increase engagement in space activities, *UPSat* was launched on 18 May 2017 [94].[34] More precisely, UPSat was the first satellite entirely developed and manufactured in Greece, that was successfully launched into orbit; it was constructed by the University of Patras and *Libre Space Foundation* [95], as part of the QB50 mission (with the ID GR-02). Importantly, UPSat was completely made of open-source software and hardware, and designed for a lifetime of 3 months [96].

Furthermore, at State level, Greece enacted law n° 4506/2017 [97] on the renewal, amendment and ratification of the contract granting the exclusive right to use—namely, a special operating license on—the radio-frequency spectrum and associated orbital slot allocated to Greece (i.e., 39 degrees East, following the granting of an exclusive license to the benefit of the Greek government). In particular, said contract stipulated that the agreement initially signed to the benefit of the company *Hellas Sat Consortium Limited* on 23 August 2001, granting it an exclusive right to use the 39 degrees East orbital slot and the radio frequencies assigned, was renewed to the benefit of the company's subsidiary. In renewing this agreement, account had been taken of the fact that the company had *inter alia* undertaken the development and launching of two new satellites—namely, *Hellas Sat 3* and *Hellas*

[34] "UPSat, the first open-source hardware and software satellite, was released in orbit by NanoRacks deployer from the International Space Station at 08:24 UTC 2017-05-18. After 30 minutes, UPSat subsystems commenced normal operations in orbit." [94].

Sat 4—to replace the existing operating satellite *Hellas Sat 2* [97][35] (in addition to upgrading and extending the *Hellas Sat* satellite system, in general).

More precisely, the two parties concerned—i.e., the Greek government on the one side, and Hellas Sat S.A. on the other—agreed for the renewal of the agreement, for an additional period of twenty years, until 14 December 2041 (through special licensing agreements signed with both the Hellenic and Cypriot Republic administrations).[36] Hence, according to law n° 4506/2017, the *Hellas Sat* operator has the obligation towards the Greek State to maintain its headquarters and self-owned premises in the territory of Greece and Cyprus [97: Art. I.δ]. Moreoever, the law requires that the satellite operation and control centre as well as backup centre can only be located in Greece and Cyprus [97: Art. I.δ and ε] and that the personnel consist solely of Greek and Cypriot nationals. Eventually, a dedicated section regarding security and confidentiality issues, allows the competent Minister to request the immediate removal of any *Hellas Sat* personnel for national security matters [97: Art. 42].

All the commitments and obligations of the beneficiary were enumerated in detail, such as to develop, implement and maintain a satellite-based telecommunication system for Greece—under the name of *Hellas Sat*–, in accordance with the provisions of the ITU rules and other obligations of the Greek State under international law, including international space law[37] (the planning and implementation phases were all evidently described in detail[38]). It is interesting to note that the company additionally committed to educate and train new corps of scientists in this field (e.g., in the field of construction, installation and operation of satellite systems and ground control).[39]

As a result, *Hellas Sat 3*—with a design lifetime of 16,5 years and carrying a payload of 47 Ku-Band transponders—was launched on 28 June 2017 [98], from Ariane Launch Complex No. 3 (ELA 3) in Kourou, French Guiana (by an Ariane 5 launch vehicle). In this regard, the satellite launch company *Arianespace* announced that "Hellas Sat 3 will deliver in-orbit back-up DTH and telecom services in its designated coverage areas, maintaining and expanding Hellas Sat business reach with additional capacity and bringing video content in High Definition and Ultra High-Definition format to the covered regions. The FSS/BSS coverage zones are Europe, the Middle East and Sub-Saharan African countries, including a cross-trap service between Europe and South Africa" [99].

[35] See, paragraph 7 of the contract preamble, and Art. 6 of the contract [97].

[36] See Part I of the contract titled "Renewal of the contract and license dated 23.08.2001", and Art. 4 of the contract.

[37] See Part II of the contract titled "Adjustment, updating and codification of the contract dated 23.08.2001", and Art. 1 of Law n° 4506/2017.

[38] *Inter alia*, Annex B contained the Implementation time-schedule; Annex C, the System Technical Specifications; Annex D, Transponders Technical specifications and delivery times and Annex E, Terminals and HUBs Technical specifications and delivery times.

[39] See Part II of the contract titled "Adjustment, updating and codification of the contract dated 23.08.2001", and Art. 18-point ζ of the Law n° 4506/2017.

Following on from that, namely a couple of years later, *Hellas Sat 4* was released on 5 February 2019 [100] (having a design lifetime of more than 15 years and carrying a payload of 34 Ku-Band transponders). Again, the satellite was launched by an Ariane 5 rocket, from the European Spaceport in Kourou, French Guiana, and programmed to "provide innovative satellite services, expanding the operations of 'Hellas Sat' in Europe, the Middle East and South Africa" [101]. More importantly, *Hellas Sat 4* was described as "one of the most advanced in the world (...) a modernized version of the well-established LM 2100 series of satellites with its successful presence in the specific market. (...) [*having*] new technology solar panels among other innovative systems. The new satellite [*did*] not have standard solar collectors that older satellites had but (...) a new sail-like set of panels that will unfold. [*A*] technology called 'multi-mission modular solar array'" [101].

To complete the picture, it should be mentioned that the implementation of the next Hellas Sat satellites (such as, *Hellas Sat 5*) is already well underway. According to the signing of an agreement that took place on 26 January 2024—creating a partnership between the European Space Agency (ESA), the French based Thales group, the National Observatory of Athens (NOA), the French space agency NCES, and with the support of the General Secretariat for Research and Innovation–, the aim is now to also "develop an optical communication payload for the upcoming new mission to be embarked on the future Hellas Sat 5 telecommunications satellite" [102].

Eventually, to abide by its obligation under international space law, Greece notified the Secretary-General of the United Nations (on the basis of a note verbale, dated 19 April 2022 [103]), about both the launching of *Hellas Sat 3* and *Hellas Sat 4* satellites.

3.2.2.2.3 Defining Key Goals in the National Strategy for Space

According to the General Secretariat of Telecommunications and Post of the Greek Ministry of Digital Governance [104], law n° 4727/2020 on Digital Governance matters [105] defined *inter alia* the key objectives of the national strategy for space. In this context, Art. 119.9 precisely highlighted that the Greek State should aim at achieving precisely defined steps and goals, through its participation in the ESA.

Hence, these goals should be: (1) to strengthen national security and defense, especially with the utilization and development of space infrastructure; i.e., to ensure national autonomy in safety and security (for border control, disaster management etc.) by enhancing the existing infrastructures (e.g., GreekCom[40]) or by developing

[40] "Since 2019 Greece has its own operational Governmental Satellite Communication (GovSatCom) system called GreeCom. The system is based on Hellas Sat 3 and 4 and the Satellite Distribution Centre (HUB) of GreeCom is installed in the Ministry of National Defense establishments. Operations are exclusively conducted by the Ministry of National Defense personnel. The system is connecting in a reliable and resilient manner decision centres (e.g President of the Republic, Prime Minister, Ministers, General Secretaries, Emergency Operation Centres (civil protection, fire brigade, coast guard), Hellenic Diplomatic Network, National Defense)" [104: 4].

new ones (e.g., small satellites), so as to autonomously respond to national safety and security needs; (2) to develop the Greek space industry; namely, to maximize the integration of Greek companies into the European industrial space landscape, to create a sustainably competitive Greek space industry; (3) to utilize space data and the development of relevant applications; such as by fostering the integration of space into society and the economy, and by facilitating the use of space technologies and applications to support public policies and business development (e.g., in telecom, transports, the maritime sector, agriculture, energy, environment etc.), the goal being to create useful public and commercial services and (4) to support space research and innovation [104: 3].

In reference to the related strategies, these should mainly serve—according to the General Secretariat of Telecommunications and Post of the Greek Ministry of Digital Governance–, purposes such as to develop and/or strengthen "(i) *a national small sat*: to achieve autonomous capabilities in order to respond to national needs, strengthening the Greek industry competitiveness with the possibility for commercial utilization in the global market; (ii) *flight heritage*: though the EU, the ESA and national programmes by maximizing technological and financial synergies; (iii) *ground infrastructure*: by utilizing and further developing existing infrastructures e.g., nationally or privately owned stations (e.g., Tanagra, Koropi, Ag. Paraskeui), telescopes, lasers, radars, etc.; (iv) *user alignment*: by supporting through space the development, implementation and monitoring of sectorial policies (e.g., telecom, transport, maritime, agriculture, energy, environment, safety and security) and (v) *data of existing infrastructure exploitation*: as the opening of public information will increase the size of the domestic market, especially in case the private and public sector work together" [104: 3].

Given this context, the Greek State adopted *inter alia* the Digital Transformation Strategy 2020-2025 [106], which mentions specific activities to be undertaken in the space domain and includes the strategy to use space as an enabler for digital transformation. Hence, particular space-related priorities were further developed in this document, like the Micro-Satellite Project [106: para. 7.1.3. titled *Projects*], GOVSATCOM—GreeCom [106: para. 7.1.3. titled *Projects*; 104][41] and the upgrade and further development of the Hellenic Copernicus Collaborative Ground Segment.[42]

[41] GOVSATCOM - GreeCom "connects the country's decision centers with voice and data services, with the aim of ensuring seamless and continuous communication of critical government services even in times of crisis. The operation of GreeCom will be expanded and supplemented with the possibilities of secure communications that will be provided by the work of microsatellites and Fiber in the sky." [104: 14].

[42] "The operation of this service which allows the free and open access as well as obtaining the data of Copernicus with the geographical coverage mentioned above, is under the responsibility of the National Observatory of Athens in cooperation with GRNET which has infrastructures and internet connectivity network of high speeds. The aim is that this computational and connectivity infrastructure will develop to an open type and not as a monolithic closed system exchanging operational existing infrastructures in particular of the European Space Agency but also National (e.g. open data for the cadastre and civil protection)." [104: 14].

As an example, the Micro-Satellite Project puts the emphasis on creating a "microsatellite constellation that will cater application for telecommunications and earth observation for their use in governmental satellite services, cartography, maritime, precision agriculture, topography, and urban planning aiming at the development of innovative integrated applications' services. [*It*] is designed to be using the Fibre in the Sky—ESA HydRON, with target the holistic provision of secure telecommunication services. Additionally, the project of microsatellites will support the application and services for search and rescue, border control, national security, civil protection. The development of the microsatellite system (space and ground) is expected to strengthen the capabilities of the Greek high-tech industry. The project is expected to increase the availability, security and authentication of governmental communication networks. In parallel, it will provide high speed connectivity in remote areas taking into consideration of CSR 3 to 2020[43]".

In conclusion, Part I focused on explaining that, from the perspective of the Greek State, involvement in space matters started in the form of specific cooperation actions in particular areas of space activity coupled with scientific collaborations, that started to be implemented since the late 80s; first *via* enhanced participation in space-related initiatives and research projects, and soon through its participation in management structures for space services, at the regional—i.e., EUTELSAT (1987), EUMETSAT (1988)—as well as international (e.g., INTELSAT, in 1988) level [107: Art. 18]. At the same time, it had acceded to four of the five principal international space law treaties, which established the legal conditions for a sustainable long-term exploration of space and the exploitation of its resources.

However, given that the participation of Greek scientists and operators in the field of space continued to increase, it soon also became clear that well-adapted rules as well as a comprehensive legal framework were eventually required. In other words, the simple transposition of international space law into national law was not sufficient any more; additional measures on the exploration and use of space resources were now necessary, to clarify the provisions of international space law treaties and to enable the Greek space actors and operators carry out space activity in an efficient and competitive manner.

References

1. Van Traa-Engelman, H. L. (1993). *Commercial utilization of outer space: Law and practice*. Martinus Nijhoff Publishers.

[43]"Additionally, the framework of development of the micro-satellites is envisaged to be part of the EU GovSatcom, EuroQCI and Connectivity announced by the European Union on 15 July 2020 and in the 23 February Action Plan on Synergies between civil, defense and space industries" [104: 6].

2. UNTC. *Convention for the establishment of a European Space Agency (with annexes and final act). Signed on 30 May 1975, entered into force 30 October 1980.* Accessed from https://treaties.un.org/Pages/showDetails.aspx?objid=08000002800df46b (ESA Convention).
3. ESA History. (2019, April 26). *A digital record of the European space pioneers.* Accessed July 2024, from https://www.esa.int/About_Us/ESA_history/A_digital_record_of_the_European_space_pioneers
4. ESA Archives Portal. (2024). *A-Z of European space. The early years: 1959-1994.* Glossary, Organizations. Accessed July 2024, from https://historicalarchives.esa.int/european-space-glossary/organisations?f%5B0%5D=glossary_az_content_title%3AE
5. Lyall, F., & Larsen, P. B. (2018). *Space law: A treatise* (2nd ed.). https://doi.org/10.4324/9781315242712
6. Krige, J., & Russo, A. (2000). *A history of the European Space Agency 1958 – 1987. Volume I The story of ESRO and ELDO, 1958 – 1973. ESA, April 2000, at 294 ("Membership of European organizations in 1965").* Accessed from https://www.esa.int/esapub/sp/sp1235/sp1235v1web.pdf
7. Mathieu, C. (2009, March 17). *Space-based services in Europe. Addressing the transition between demonstration and operation. ESPI report.* Accessed from https://www.espi.or.at/wp-content/uploads/espidocs/Public%20ESPI%20Reports/espi_report_17.pdf
8. EUTELSAT GROUP. (n.d.). *Our history.* Accessed July 2024, from https://www.eutelsat.com/en/group/our-history.html
9. EUMETSAT. (n.d.). *About us. Who we are.* Accessed July 2024, from https://www.eumetsat.int/about-us/who-we-are
10. EUMETSAT.INT. (n.d.). *The European Space Agency (ESA) and EUMETSAT have a long-standing relationship.* Accessed July 2024, from https://www.eumetsat.int/european-space-agency-esa
11. ESA. (2019, April 26). *A digital record of the European space pioneers.* Accessed July 2024, from https://www.esa.int/About_Us/ESA_history/A_digital_record_of_the_European_space_pioneers
12. ESA Archives Portal. (2024). *A-Z of European space. The early years: 1959-1994. Glossary, projects.* Accessed July 2024, from https://historicalarchives.esa.int/european-space-glossary/projects
13. EUTELSAT IGO. *Background.* Accessed July 2024, from https://www.eutelsatigo.int/en/about/
14. UNOOSA. *Convention establishing the European Telecommunications Satellite Organization "EUTELSAT" (with annexes, operating agreement and protocol of amendment of 15 December 1983), 15 July 1982, UNTS 1519 (p. 149), entered into force 1 September 1985.*
15. United Nations Treaty Collection (UNTC). (n.d.). *Convention establishing the European Telecommunications Satellite Organization "EUTELSAT" (with annexes, operating agreement and protocol of amendment of 15 December 1983) – Ratifications.* Accessed July 2024, from https://treaties.un.org/Pages/showDetails.aspx?objid=08000002800c637e
16. *Law n° 1713/1987 on the ratification of the convention and agreement pertaining to the creation and functioning of the European Telecommunications Satellite Organization (EUTELSAT), together with its annexes and protocols.* Published at the Greek Government Gazette A' 116/08.07.1987, p. 1141 et seq.
17. Treaty Series 2009. *Protocol on the privileges and immunities of the European Telecommunications Satellite Organization (EUTELSAT), 13 February 1987, UNTS v. 1512 (p. 11), entered into force 17 August 1988.*
18. *Law n° 2280/1995 on the ratification of the protocol on the privileges and immunities of the European Telecommunications Satellite Organization (EUTELSAT).* Published at the Greek Government Gazette A' 10/26.01.1995, p. 99 et seq.
19. EUTELSAT IGO. *Law n° 2962/2001 on the ratification of the amended convention on the establishment of the European Telecommunications Satellite Organization (EUTELSAT).* Published at the Greek Government Gazette A' 267/23.11.2001, p. 3751 et seq.

20. EUTELSAT IGO. *Governance*. Accessed July 2024, from https://www.eutelsatigo.int/en/governance/
21. EUTELSAT IGO. *Member states*. https://www.eutelsatigo.int/en/member-states/
22. ESA Archives Portal. (2024). *A-Z of European space. The early years: 1959-1994. Glossary, projects*. Accessed July 2024, from https://historicalarchives.esa.int/european-space-glossary/projects?page=2
23. EUMETSAT.INT. (n.d.). *Meteosat first generation (retired)*. Accessed July 2024, from https://www.eumetsat.int/meteosat-first-generation-retired
24. EUMETSAT. (2011). *EUMETSAT 25 years. Foundations for the future*. EUMETSAT. ISBN 978-92-9110-091-0. Accessed from https://www-cdn.eumetsat.int/files/2020-04/pdf_br_history-book.pdf
25. EUMETSAT. *Convention for the establishment of a European Organization for the Exploitation of Meteorological Satellites (EUMETSAT), entered into force on 19 June 1986, including amendments entered into force on 19 November 2000*. Accessed July 2024, from https://www-cdn.eumetsat.int/files/2020-04/pdf_leg_convention.pdf
26. EUMETSAT. *Law 1770/12.04.1988 on the ratification of the convention establishing the European Organization for the exploitation of Meteorological Satellites*. Published at the Greek Government Gazette A' 69, p. 1661 et seq.
27. EUMETSAT. *Law 3019/13.06.2002 on the ratification of the protocol on privileges and immunities of the European Organization for the exploitation of Meteorological Satellites (EUMETSAT)*. Published at the Greek Government Gazette A' 141/19.06.2002, p. 2805 et seq.
28. EUMETSAT. *Protocol on the privileges and immunities of the European Organization for the Exploitation of Meteorological Satellites (EUMETSAT) entered into force on 5 January 1989, as amended in accordance with notification of the depositary of the protocol on privileges and immunities dated 12 January 2004*. Accessed July 2024, from https://www.eumetsat.int/media/40626
29. *Presidential Decree 161/20.06.1997*. Published at the Greek Government Gazette A' 142, p. 6203.
30. EUMETSAT.INT. (n.d.). *Who we are*. Member States. Accessed July 2024, from https://www.eumetsat.int/who-we-are/eumetsat-member-states
31. EUMETSAT.INT. *EUMETSAT annual report 2023*, p. 104. Accessed July 2024, from https://www.eumetsat.int/media/52100
32. ESA. (2014). *About us*. ESA History. Accessed July 2024, from https://www.esa.int/About_Us/ESA_history/The_ESRO_Convention_and_Juste_retour
33. ESA. *Enabling & support*. Space engineering & Technology. Accessed July 2024, from https://www.esa.int/Enabling_Support/Space_Engineering_Technology/European_Space_Agency_ESA
34. ESA. (2005, March 22). *Greece becomes 16th ESA Member State*. Accessed July 2024, from https://www.esa.int/About_Us/Business_with_ESA/Greece_becomes_16th_ESA_Member_State
35. ESA. *N° 20–1994: ESA/Greece Cooperation Agreement (4 July 1994)*. Accessed July 2024, from https://www.esa.int/Newsroom/Press_Releases/ESA_Greece_Cooperation_Agreement
36. *Law n° 2953/19.10.2001 on the ratification of the agreement between the government of the Hellenic Republic and the European Space Agency concerning space cooperation for peaceful purposes*. Published at the Greek Government Gazette A' 249/22.10.2001, p. 3479. et seq.
37. *Law n° 3308/28.01.2005 on the ratification of the agreement between the government of the Hellenic Republic and the European Space Agency pertaining to the accession of Greece to the convention for a European Space Agency and the associated terms and conditions*. Published at the Greek Government Gazette A' 21/01.02.2005, p. 117. et seq.
38. European Student Think Tank. (2019) *The participation of Greece in the European Space Agency (ESA): An interview with an ex-delegate of Greece to ESA*. Accessed July 2023, from https://esthinktank.com/the-participation-of-greece-in-the-european-space-agency-esa-an-interview-with-an-ex-delegate-of-greece-to-esa/

39. Klock, E., & Aliberti, M. *ESA enlargement. What interested countries can do to prepare themselves for ultimate accession – With a special focus on the CEE Region, ESPI Report 47 January 2014*. Accessed July 2024, from https://www.files.ethz.ch/isn/176962/ESPI_Report_47.pdf
40. ESA. *Greece to sign agreement on accession to ESA Convention (22.06.2024)*. Accessed July 2024, from https://www.esa.int/About_Us/Business_with_ESA/Greece_to_sign_agreement_on_accession_to_ESA_Convention
41. *Explanatory Memorandum to the Draft law entitled Authorization of space activities - Registration in the National Registry of Space Objects - Establishment of the Hellenic Space Agency and other provisions*. Accessed from https://www.hellenicparliament.gr/Nomothetiko-Ergo/Anazitisi-Nomothetikou-Ergou?law_id=ca706be4-0ccc-4dbf-a8ce-a83b01095b01 (in Greek).
42. ESA. *Greece becomes 16th ESA Member State (22.03.2005)*. Accessed July 2024, from https://www.esa.int/About_Us/Business_with_ESA/Greece_becomes_16th_ESA_Member_State
43. Greek Parliament. *Explanatory memorandum to the draft law Ratification of the agreement between the Member States to the convention for the establishment of the European Space Agency and of the European Space Agency for the protection and exchange of classified information*. Accessed July 2024, from https://www.hellenicparliament.gr/UserFiles/2f026f42-950c-4efc-b950-340c4fb76a24/%CE%88%CE%B3%CE%B3%CF%81%CE%B1%CF%86%CE%BF%20(K_SPACE_EIS).pdf
44. UNOOSA. *COPUOS, Fifty-seventh session, Vienna, 9–20 April 2018, 'The European Space Agency as a mechanism and an actor of international cooperation', A/AC.105/C.2/2018/CRP.20*.
45. Masson-Zwaan, T. L. (2010). Recent developments in EU space policy and law. *The Korean Journal of Air and Space Law, 25*(2), 231–247. Accessed from https://hdl.handle.net/1887/17593
46. EUR-Lex. *Green Paper - European Space Policy/* COM/2003/0017* final */. Accessed from https://eur-lex.europa.eu/legal-content/EN/TXT/HTML/?uri=CELEX:52003DC0017
47. ESA. (2001, February 01). *Greece signs special accord with ESA*. Accessed July 2024, from https://www.esa.int/About_Us/Corporate_news/Greece_signs_special_accord_with_ESA
48. ESA. (2007, September 28). *Athens workshop on space industry*. Accessed July 2024, from https://www.esa.int/About_Us/Business_with_ESA/Athens_workshop_on_space_industry
49. ESA bulletin 120 - November 2004. *Enlarging ESA? – After the accession of Luxembourg and Greece*. Accessed from https://www.esa.int/esapub/bulletin/bulletin120/bul120g_poncelet.pdf
50. ESA. (n.d.). *Greek ESA BIC opens in a new space era*. Accessed July 2024, from https://business.esa.int/news/greek-esa-bic-opens-new-space-era
51. ESA. (2023, June 20). *ESA backs Greek firms' and universities' CubeSats*. Accessed July 2024, from https://www.esa.int/Applications/Connectivity_and_Secure_Communications/ESA_backs_Greek_firms_and_universities_CubeSats
52. Corallia. *Greek space catalogue, 2019*. Accessed from https://www.hellenic-asi.org/files/si-cluster.pdf
53. ESA. (2014). *General Secretariat for Research and Technology (GRST)—European Space Agency (ESA), Greece Space Capabilities*. Accessed from https://helas.gr/esa/Greek_Space_Catalogue.pdf
54. ESA. *Funding (for 2024)*. Accessed July 2024, from https://www.esa.int/About_Us/Corporate_news/Funding
55. von der Dunk, F. (2006). Fundamental provisions for National Space Laws. In *Meeting international responsibilities and addressing domestic needs: Proceedings United Nations, Vienna, 2006*. Accessed from https://digitalcommons.unl.edu/cgi/viewcontent.cgi?article=1010&context=spacelaw

56. Hellenic Republic, Ministry of Foreign Affairs. *Greece's course in the EU*. Accessed July 2024, from https://www.mfa.gr/en/foreign-policy/greece-in-the-eu/greeces-course-in-the-eu.html
57. Matxalen Sanchez Aranzamendi. (2009, September 21). *Economic and policy aspects of space regulations in Europe. Part II: The case of national space legislation – Finding the way between common and coordinated action, ESPI, Report*, p. 36. Accessed from https://www.files.ethz.ch/isn/124779/espi%20report%2021.pdf
58. Sheehan, M. (2011). EU strategy, CSDP and the European Space Programme. *Studia Diplomatica, 64*(1), 41–54. Accessed from https://www.jstor.org/stable/10.2307/26531481
59. *Parliament Resolution, 17 January 2002, PR TAPROV (2002)0015 «Europe and Space»*. Accessed from https://www.europarl.europa.eu/doceo/document/TA-5-2002-0015_EN.pdf?redirect
60. European Commission. *White paper Space: a new European frontier for an expanding Union An action plan for implementing the European Space policy {SEC (2003) 1249 } /* COM/2003/0673 final */*. Accessed from https://eur-lex.europa.eu/legal-content/EN/TXT/?uri=celex%3A52003DC0673
61. European Commission. *Green Paper European Space Policy [Brussels, 21.1.2003 - COM (2003) 17 final]*. Accessed from https://eur-lex.europa.eu/legal-content/EN/TXT/PDF/?uri=CELEX:52003DC0017
62. European military satellites: New tools for defence cooperation. (2000). *Strategic Comments, 6*(10), 1–2. https://doi.org/10.1080/1356788000602
63. Athens Exchange Group (ATHEX GROUP). (n.d.). *Press release Hellenic Telecommunications Organization S.A.* Accessed July 2024, from https://www.athexgroup.gr/permalink/-/asset_publisher/contentdispl/content/press-release-hellenic-telecommunications-organization-s-a-3106
64. N-takosnews. (2021, August 31). *Greece in ... space*. Accessed July 2024, from https://n-takosnews.gr/i-ellada-sto-diastima/
65. Iordanou, H., & Protopapas, C. *HELLAS SAT: The birth of a new satellite operator*. Accessed July 2024, from https://arc.aiaa.org/doi/pdf/10.2514/6.2004-364-204
66. EUR-Lex. *Council Resolution of 16 November 2000 on a European Space Strategy (2000/C 371/02)*. Accessed from https://www.copernicus.eu/sites/default/files/2018-10/2000c_371_2_space_strategy.pdf
67. EUR-Lex. (2007, December 17). Treaty of Lisbon amending the Treaty on European Union and the Treaty establishing the European Community (Lisbon 13 December 2007). *Official Journal C 50*, 306. Accessed from https://eur-lex.europa.eu/legal-content/EN/TXT/PDF/?uri=OJ:C:2007:306:FULL
68. EUR-Lex. (2003, July 18). Draft treaty establishing a constitution for Europe. *Official Journal C 169, at 0001 – 0105*. Accessed from https://eur-lex.europa.eu/legal-content/EN/TXT/HTML/?uri=CELEX:52003XX0718(01)&rid=1
69. von der Dunk, F. G. (2018). The European Union and space—Space for competition? In *Proceedings of the 61st (2018) colloquium on the law of outer space*, vol. 61, pp. 285–300.
70. Robinson J., & Romancov, M. (2014, January). *The European Union and space: Opportunities and risks. EU non-proliferation consortium, non-proliferation papers - no. 37*. Accessed from https://www.sipri.org/sites/default/files/EUNPC_no-37-1.pdf
71. Johnson, C. *Draft international code of conduct for outer space activities, fact sheet*. SWF. Accessed from https://swfound.org/media/166384/swf_draft_international_code_of_conduct_for_outer_space_activities_fact_sheet_february_2014.pdf
72. Jinyuan, S., & Lixin, Z. (2014). The European Union draft code of conduct for outer space activities: An appraisal. *Space Policy, 30*, 34–39. Accessed from https://www.sciencedirect.com/science/article/abs/pii/S0265964614000034
73. EEAS. *Draft International Code of conduct for outer space activities – Version 31 March 2014*. Accessed July 2024, from https://eeas.europa.eu/archives/docs/non-proliferation-and-disarmament/pdf/space_code_conduct_draft_vers_31-march-2014_en.pdf

74. EUR-Lex. Framework agreement between the European Community and the European Space Agency. *OJEU L.261/64 – 6.8.2004*. Accessed from https://eur-lex.europa.eu/resource.html?uri=cellar:742587dd-44f3-4527-be91-a2eab676330f.0008.02/DOC_2&format=PDF
75. ESA. (2004, November 25). *First ever 'space council' paves the way for a European space programme*. Accessed July 2024, from https://www.esa.int/About_Us/Corporate_news/First_ever_Space_Council_paves_the_way_for_a_European_space_programme
76. ESA. (2005, November 28). *Global monitoring for environment and security is main issue for Third Space Council*. Accessed July 2024, from https://www.esa.int/About_Us/Corporate_news/Global_Monitoring_for_Environment_and_Security_is_main_issue_for_Third_Space_Council
77. ESA. (2008, September 26). *Ministers meet to take forward the European Space Policy (5th Space Council)*. Accessed from https://www.esa.int/About_Us/Corporate_news/Ministers_meet_to_take_forward_the_European_Space_Policy_5th_Space_Council
78. ESA. (2009, May 29). *Space ministers highlight contribution of space to leading economic recovery through innovation*. Accessed from https://www.esa.int/About_Us/Law_at_ESA/Space_ministers_highlight_contribution_of_space_to_leading_economic_recovery_through_innovation
79. ESA. (2010, December 07). *Seventh Space Council supports strong European leadership in space*. Accessed from https://www.esa.int/About_Us/Corporate_news/Seventh_Space_Council_supports_strong_European_leadership_in_space
80. ESA. (2007, May 22). *Europe's Space Policy becomes a reality today (4th Space Council)*. Accessed from https://www.esa.int/About_Us/Corporate_news/Europe_s_Space_Policy_becomes_a_reality_today_4th_Space_Council
81. ESA. *European Commission-ESA, Resolution on the European Space Policy, ESA BR 269 22.05.07*. Accessed July 2024, from https://esamultimedia.esa.int/docs/BR/ESA_BR_269_22-05-07.pdf
82. ESA. *EU/ESA Space Council*. Accessed July 2024, from https://www.esa.int/About_Us/Corporate_news/EU_ESA_Space_Council
83. ESA. (2021, November 19). *N° 39–2021: Decisions from the Intermediate Ministerial Meeting 2021*. Accessed July 2024, from https://www.esa.int/Newsroom/Press_Releases/Decisions_from_the_Intermediate_Ministerial_Meeting_2021
84. UN Secretariat COPUOS. (2008). *Information furnished in conformity with the convention on registration of objects launched into outer space. Note verbale dated 25 March 2004 from the permanent mission of Greece to the United Nations addressed to the secretary-general*. 29 March 2004. UN DOC ST/SG/SER.E/446*. Accessed from https://www.unoosa.org/documents/pdf/ser446E.pdf
85. Papadimitriou, A., Adriaensen, M., Antoni, N., & Giannopapa, C. (2019). Perspective on space and security policy, programmes and governance in Europe. *Acta Astronaumica, 161*, 183–191.
86. *Law n° 3546/2007 on the ratification of the collaboration agreement between the defense minister of Greece and the defense ministers of France, Belgium, Spain and Italy pertaining to the participation of the Hellenic Republic in the cooperation between France, Belgium, Spain and Italy for the Earth Observation space system HELIOS II*. Published at the Greek Government Gazette A' 66/20.03.2007, p. 1663 et seq.
87. SpaceNews. (2020, January 09). *2010-2019: The decade in space*. Accessed July 2024, from https://spacenews.com/2010-2019-the-decade-in-space/
88. European Commission. *Communication from the Commission to the European Parliament, the Council, the European Economic and Social Committee and the Committee of the Regions - Space Strategy for Europe*. Brussels 26.10.2016 - COM(2016) 705 final. Accessed August 2023, from eur-lex.europa.eu/legal-content/EN/TXT/PDF/?uri=CELEX:52016DC0705&from=EN
89. Evroux, C. (2022). *EU space policy: Boosting EU competitiveness and accelerating the twin ecological and digital transition. EPRS | European Parliamentary Research Service, Briefing*

PE 698.926. Accessed from https://www.europarl.europa.eu/thinktank/en/document/EPRS_BRI(2022)698926

90. EUR-Lex. *Regulation (EU) 2021/696 of the European Parliament and of the Council of 28 April 2021 establishing the Union Space Programme and the European Union Agency for the Space Programme and repealing Regulations (EU) No 912/2010, (EU) No 1285/2013 and (EU) No 377/2014 and Decision No 541/2014/EU*. Accessed from https://eur-lex.europa.eu/eli/reg/2021/696/oj
91. Hellenic Republic, Ministry of Digital Governance - General Secretariat of Telecommunications and Post. (2020). *Greek initiative announcement of opportunity under the ESA ARTES Space Systems for Safety & Security (4S) Strategic Programme Line*. Accessed July 2024, from https://www.google.com/url?sa=t&rct=j&q=&esrc=s&source=web&cd=&cad=rja&uact=8&ved=2ahUKEwjTrfK5x8WAAxW2_rsIHf4bBdMQFnoECA0QAQ&url=https%3A%2F%2Fmindigital.gr%2Fwp-content%2Fuploads%2F2020%2F11%2F4S-GREEK-INITIATIVE-THROUGH-ARTES-4S-GR.docx&usg=AOvVaw0PSQ94vkpoPwyQTNmsNXsP&opi=89978449
92. Agence Europe. (2021, June 22). *Europe Daily Bulletin No. 12746. European Commission and European Space Agency officially sign Financial Framework Partnership Agreement*. Accessed July 2024, from https://agenceurope.eu/en/bulletin/article/12746/17
93. *Law n° 4508/2017 on the Authorization of space activities - Registration in the National Registry of Space Objects - Establishment of the Hellenic Space Agency and other provisions*. Published at the Greek Government Gazette A' 200/22.12.2017, p. 4411 et seq.
94. Libre Space Foundation. (n.d.). *Successful deployment of UPS at, the first open source satellite*. Accessed July 2024, from https://libre.space/2017/05/19/upsat-successful-deployment/
95. Libre Space Foundation. (n.d.). *About us*. Accessed July 2024, from https://libre.space/about-us/
96. Libre Space Foundation. (n.d.). *UPSat (QB50 GR02)*. Accessed July 2024, from https://space.skyrocket.de/doc_sdat/upsat.htm
97. *Law n° 4506/2017 on the Renewal - Amendment - Codification of the Agreement on the Granting of a Special Exploitation License on the exclusive right of the Greek State to access and use the nominal position of 39° east on the orbit of the earth's artificial geostationary satellites and the associated telecommunication radio frequencies for the purpose of construction, launch, operation and commercial exploitation of a pan-Hellenic and cross-border satellite telecommunications system under the name HELLAS SAT and other provisions*. Published at the Greek Government Gazette A' 191/12.12.2017, p. 4255 et seq.
98. H-hellas sat. (n.d.). *Hellas Sat 3*. Accessed July 2024, from https://www.hellas-sat.net/hellas-sat-3/
99. arianespace arianegroup. (2017, June 22). *Arianespace to launch on June 28 Hellas Sat 3-Inmarsat S EAN for Inmarsat and Hellas Sat; and GSAT-17 for the Indian Space Agency ISRO*. Accessed July 2024, from https://www.arianespace.com/press-release/arianespace-to-launch-on-june-28-hellas-sat-3-inmarsat-s-ean-for-inmarsat-and-hellas-sat-and-gsat-17-for-the-indian-space-agency-isro/
100. H-hellas sat. (n.d.). *Hellas Sat 4*. Accessed July 2024, from https://www.hellas-sat.net/hellas-sat-4
101. New Greek Television. (2019, February 06). *"Hellas Sat 4" launched (video), by E. Tsiliopoulos*. Accessed July 2024, from https://www.newgreektv.com/news-in-english-for-greeks/greece/item/27733-hellas-sat-4-launched-video
102. OT English Edition. (2024, February 01). *Internet Speed in Greece to surge 1,000 times with Hellas Sat 5*. Accessed July 2024, from https://www.ot.gr/2024/02/01/english-edition/internet-speed-in-greece-to-surge-1000-times-with-hellas-sat-5/
103. UN Secretariat. *Information furnished in conformity with the convention on registration of objects launched into outer space - Note verbale dated 2 March 2022 from the permanent mission of Greece to the United Nations (Vienna) addressed to the Secretary-General*

(19 April 2022) UN Doc ST/SG/SER.E/1044. Accessed from https://www.unoosa.org/oosa/en/osoindex/data/documents/gr/st/stsgser.e1044.html

104. Hellenic Republic, Ministry of Digital Governance - General Secretariat of Telecommunications and Post. (2021). *Request for Information (RFI) - Outline concept for end-to-end small satellites multipurpose solutions in response to national and European needs* - date of issue 28.06.2021 (*available online*).
105. *Law n° 4727/2020 on the Digital Governance (Incorporation into Greek Law of Directive (EE) 2016/2102 and Directive (EE) 2019/1024) – Electronic Communications (Incorporation into Greek Law of the Directive (EE) 2018/1972) and other provisions*. Published at the Greek Government Gazette A' 184/23.09.2020, p. 9623 et seq.
106. *Decision of the Minister of State and Minister of Digital Governance n° 120301 ΕΞ 2021/18.06.2021 on the Digital Transformation Bible*. Published at the Greek Government Gazette B' 2894/05.07.2021, p. 37711 et seq.
107. Ministry of Digital Governance. *Public consultation procedure - Comments under draft law n° 4508/2017 (in Greek)*. Accessed July 2024, from http://www.opengov.gr/digitalandbrief/?p=771

Part II
Greek Space Law

Following the adoption—a few decades ago—of the five UN international space law treaties,[1] a set of ground rules were set up for the conduct of space activity. However, these legal instruments primarily contained general norms and conditions applicable to such activities; hence, most of them had to be supplemented by more specific measures, to better address particular or new aspects of space operations.

Be that as it may, no further binding agreement has been adopted up to now—i.e., either through the UNCOPUOS or *via* other international space law *fora*—since 1979, when the last treaty (namely, the Moon Agreement) was adopted [2]. Put differently, practice showed that reaching international consensus on a legally binding agreement for the regulation of specific (or emerging) aspects of space activity is proving difficult; this opened the door for the adoption of a wide range of soft-law instruments (e.g., UN Resolutions) to elaborate and interpret rules of international space law, as well as to address new challenges facing the space sector. In this context, although national legislators were not *stricto sensu* obligated to adopt domestic laws implementing (or complementing) the international space law treaties, such initiatives were in truth more than necessary.

More precisely, States were virtually required—by the demands of the situation—to adopt measures in national law, to *first* implement Art. VI, VII and VIII of the Outer Space Treaty (OST) laying down the rules and basic principles governing issues of responsibility, liability and registration (in conjunction with the Liability Convention and the Registration Convention) [3: 5]. Therefore, in numerous States, additional clear and targeted measures were adopted to regulate the practical aspects of space activities undertaken at the national level, e.g., to better organize the authorization procedure allowing space operators to carry out space activities; the functioning and competences of the licensing authority; the operation and supervision of space objects; or measures aimed at protecting the outer space

[1] Namely, the Outer Space Treaty (OST)—also referred to as the 'Constitution' of International Space Law [1: 377]—, the Rescue Agreement, the Liability Convention, the Registration Convention and the Moon Agreement.

environment. *Second*, along with the adoption of substantive rules of law on the adequate and sustainable realization of space operations, competent bodies and authorities being assigned specific powers and tasks in reference to space matters had also to be created, i.e., for the purpose of an efficient coordination and supervision.

Furthermore, to provide help in the adoption of national space legislations, a series of soft-law instruments were enacted; namely UN Resolutions on related issues [3; 1 at 386] such as the 2004 Resolution on the Application of the Concept of the 'Launching State' [4]; the 2007 Resolution on the Recommendations on Enhancing the Practice of States and International Intergovernmental Organizations in Registering Space Objects [5]; the 2007 Space Debris Mitigation Guidelines of the Committee on the Peaceful Uses of Outer Space [6] (later endorsed by the UNGA in its Resolution 62/217) [7] and the 2013 Resolution on Recommendations on National Legislation Relevant to the Peaceful Exploration and Use of Outer Space [8].

Finally, text building-blocks were proposed with the same purpose—that is, to assist national legislators in the drafting of a coherent regulatory framework for space activities—, during the *Workshop on National Space Legislation* held in 2004, in the context of Project 2001 Plus [9].[2] These building blocks were then used by the ILA as a basis for the preparation of a *Draft Model Law*, released during its 75th Conference held on 26–30 August 2012 in Sofia, as resolution 6/2012 [9]. The *Sofia Guidelines for a Model Law on National Space Legislation*—referred to as the *Sofia Guidelines*—, contained draft rules regulating all the principal space activity topics. Hence, although not mandatory, they provided valuable advice to be taken into account by many States in the drafting of their national space laws and policies.

Recognizing that participating in space activity required a further regulatory effort, implementing international space law and at the same time promoting its strategic priority areas, the Greek State adopted its national space law in 2017, with the aim to establish a competitive and well adapted legal framework. The legal instrument that was enacted to this end is law n° 4508/2017 (also law n° 4508/2017 or the Greek space law), entitled **Authorization of space activities - Registration in the National Registry of Space Objects - Establishment of the Hellenic Space Agency and other provisions** [10].[3]

For a better understanding of the Greek space law framework, it is essential to note that useful clarifications are being provided by the **Explanatory**

[2] The *Workshop on National Space Legislation*, held 2004 in the framework of Project 2001 Plus, was a cooperation exercise between the Cologne Institute of Air and Space Law and the German Aerospace Centre (DLR). The building-blocks referred *inter alia* to authorization and supervision of space activities; registration of space objects; compensation; regulation and additional regulation [9].

[3] The content of law n° 4508/2017 is not strictly limited to space activity and space matters; namely, the whole document comprises a total of 60 articles, where only articles 1 to 24 regulate space activity, the other ones referring to different issues. Therefore, Articles 25 to 59 of law n° 4508/2017 are irrelevant to space activity [10].

Memorandum to the draft of the Greek space law, as mentioned in this Part of the study where appropriate (hereafter, Explanatory Memorandum) [11]; especially given that said instrument refers to specific legal instruments, that must be taken into account as a supplement to national measures. Likewise, it is reasonable to also consider that rules or instruments released after the enactment of the Greek space law, i.e. updating previously adopted norms, have to be considered; indeed, the 2019 COPUOS *Long-Term Sustainability Guidelines* [12] explicitly encourage national legislators to adopt and keep up-to-date their domestic space laws.[4]

In this context, the provisions which are regulating space matters in the Greek space law appear to be divided into two distinct areas, as follows: a first category of rules (i.e., comprising articles 1 to 17) regulates the conduct of space activities, while a second one (containing articles 18 to 24) provided for the creation of the Greek space authority. A more detailed analysis shows that law n° 4508/2017 has eventually—more-or-less—followed the Sofia Guidelines; as a result, the Greek space law provides a general, well-articulated and comprehensive framework, rather than a sector-based approach that would address space matters by different acts or regulations, unlike other States [3: 3]. Accordingly, it is applicable to all space activities and space operators that are under the responsibility of the Greek government, in accordance with Art. VI, VII and VIII of the OST.

The present study will follow this two-pronged perspective, but it will *first* present (i.e., in this Part II) the Greek State and non-State authorities and bodies, competent to adopt any measures required to implement the provisions of the international space law treaties (Chap. 4). Following on from that, focus will be put on the rules of substantive law applicable to space activity and space operators, as principally laid down in law n° 4508/2017 (Chap. 5).

References

[1] Handmer, A., & Freeland, S. (2022). The use of law to address space debris mitigation and remediation: Looking through a science and technology lens. *Journal of Air Law and Commerce, 87*, 375. https://doi.org/10.25172/jalc.87.3.2

[2] Jakhu, R. S., & Freeland, S. R. (2016). The relationship between the outer space treaty and customary international law. In *Proceedings of the 67th International Astronautical Congress (IAC 2016): Making space accessible and affordable to all countries* (pp. 11648–11659). International Astronautical Federation, IAF. Accessed from https://iafastro.directory/iac/archive/browse/IAC-16/E7/2/32294

[3] Froehlich, A., & Seffinga, S. (Eds.). (2018). *National space legislation – A comparative and evaluative analysis*. Springer.

[4] United Nations General Assembly. (2005). *Resolution 59/115: Application of the concept of the "launching State"*. UNGA 59th Session, UN Doc. A/RES/59/115.

[4] In particular, the *Long-Term Sustainability Guidelines* provides that national legislators should "Adopt, revise and amend, as necessary, national regulatory frameworks for outer space activities" [12: Guideline A.1] and "Consider a number of elements when developing, revising or amending, as necessary, national regulatory frameworks for outer space activities" [12: Guideline A.2].

Retrieved from https://undocs.org/en/A/RES/59/115 (Resolution on the Concept of 'Launching State').
[5] United Nations General Assembly. (2008). *Resolution 62/101: Recommendations on enhancing the practice of states and international intergovernmental organizations in registering space objects*. UNGA 62nd Session, UN Doc. A/RES/62/101. Retrieved from https://www.unoosa.org/documents/resolutions/a/res62_101E.pdf (Resolution on Enhancing Registration).
[6] United Nations. (2007). *Report of the United Nations Committee on the Peaceful Uses of Outer Space*. Official records of the General Assembly, Sixty-second Session, Supplement No. 20 (A/62/20). UN Doc. A/62/20 – Annex, p. 47. Accessed from https://www.unoosa.org/pdf/gadocs/A_62_20E.pdf
[7] General Assembly Resolution. *62/217 International cooperation in the peaceful uses of outer space*. UNGA Sixty-second session. UN Doc A/RES/62/217* (1 February 2008). https://www.unoosa.org/pdf/gares/ARES_62_217E.pdf
[8] United Nations General Assembly. (2008). *Resolution 62/217: International cooperation in the peaceful uses of outer space*. UN Doc. A/RES/62/217. Accessed from https://www.unoosa.org/pdf/gares/A_RES_68_074E.pdf
[9] UN COPUOS LSC. (2013, March 26). *Information on the activities of international intergovernmental and non-governmental organizations relating to space law: Note by the Secretariat*. UN Doc A /AC.105/C.2/2013/CRP.6. Accessed from https://www.unoosa.org/pdf/limited/c2/AC105_C2_2013_CRP06E.pdf
[10] *Law n° 4508/2017 on the Authorization of space activities - Registration in the National Registry of Space Objects - Establishment of the Hellenic Space Agency and other provisions*. Published at the Greek Government Gazette A' 200/22.12.2017, p. 4411 et seq.
[11] *Explanatory Memorandum to the draft law on the authorization of space activities - Registration in the National Registry of Space Objects - Establishment of the Hellenic Space Agency and other provisions*. Accessed from https://www.hellenicparliament.gr/Nomothetiko-Ergo/Anazitisi-Nomothetikou-Ergou?law_id=ca706be4-0ccc-4dbf-a8ce-a83b01095b01 (in Greek).
[12] United Nations. (2019). *Report of the United Nations Committee on the Peaceful Uses of Outer Space*. Official Records of the General Assembly, Seventy-fourth Session, Supplement No. 20 (A/74/20). UN Doc. A/74/20, p. 54. Accessed from https://www.unoosa.org/res/oosadoc/data/documents/2019/a/a7420_0_html/V1906077.pdf

Chapter 4
Competent State and Non-State Greek Authorities

The Greek space law (law n° 4508/2017) created a national authority responsible to decide on space matters, initially under the name of *Hellenic Space Agency S.A.* and later reshaped as *Hellenic Space Center* (i.e., following adoption of law n° 4623/2019), with functions including powers of supervision and control. However, since other State and non-State bodies are also competent to decide on space issues, the following paragraphs provide an overview of the principal authorities for a better understanding.

As a preliminary remark, it should be noted that all national measures and laws must be debated in the Greek Parliament, and enacted in line with the relevant articles of the Greek Constitution, as currently in force.[1] More precisely, according to Art. 70 para. 1 of the Greek Constitution, the Parliament conducts its main activity in Plenum; to steamline its work, Standing Parliamentary Committees are set up at the beginning of each regular session, for the examination and procession of bills and law proposals that will be submitted [1: 68 para. 1].[2]

In this context, Art. 72 para. 1 of the Greek Constitution enumerates which topics are within the exclusive legislative competence of the Plenum, and a careful reading of said article reveals that space matters are not explicitly mentioned *per se*.[3] Be that

[1] The Greek Constitution was revised by the Parliamentary Resolution of November 25, 2019 of the IXth Revisionary Parliament [1].

[2] "At the beginning of each regular session, Parliament shall set up standing parliamentary committees composed of Members of Parliament for the examination and processing of Bills and law proposals submitted, as specified by the Standing Orders of the Parliament". Hence, the preliminary work for the Greek Parliament is conducted by such standing parliamentary committees, as specified by the Standing Orders and subject to Art. 72 of the Constitution defining the exclusive competence of the Plenum. The standing parliamentary committees are empowered with legislative responsibilities, by taking part in the process of debating and voting on matters which do not fall within the exclusive competence of the Plenum [2].

[3] According to Art. 72 para. 1 of the Greek Constitution, the Plenum exercises exclusive competence over such matters as those referred to in Art. 3 (relations of Church and State); Art. 13 (freedom of religious conscience); Art. 27 (changes in the State's boundaries and presence of foreign military

A. Koskina, M. Plionis, *Greek Space Law*, Space and Society,
https://doi.org/10.1007/978-3-031-72813-6_4

as it may, in line with this provision, the Plenum is competent for the ratification of *any* international law treaty or agreement, which therefore includes space law treaties. At the same time, it is also empowered to decide on any threat to national security or to domestic interests, or in the event that an activity has a potential effect on the exercise and protection of individual rights; all of which could well arise from the uses of space resources or space systems. Further features of space matters may be addressed within other parliamentary organs, such as the competent Standing Parliamentary Committees [1: Art. 70 para. 2, coupled with Art. 72 para. 3 laying down that the Standing Orders determine the allocation of competences among ministries to these committees], as well as through relevant initiatives adopted by ministries in areas under their competence, and in the context of the national general policy [2].

As a result, an equally important role for the regulation of space matters is played by the competent ministry, which has more practical knowledge and experience of the conditions in which space activity is being carried out. Following this reasoning, this study will *first* clarify the tasks and competences within the competent ministry (Sect. 4.1), and then present the Greek national authority for space matters (Sect. 4.2) as well as other space-related entities (Sect. 4.3).

4.1 The Ministry of Digital Governance

In the Greek framework, the Ministry of Digital Governance [*Υπουργείο Ψηφιακής Διακυβέρνησης*] is the authority currently empowered to prepare and initiate legislation on space matters, as well as to implement space laws. It is at the same time responsible for the introduction of measures aimed at promoting space activity, with the general objective of also strengthening and protecting the interests of Greek space stakeholders.

In this context, Art. 13 of Law n° 4622/2019 [3] as it is currently in force, stipulates that all ministers have, in line with the Constitution and the Parliament Standing Orders, the right of legislative initiative over any matters within their competences and in the context of the governmental policy [3: Art. 13.γ].[4] In

forces); Art. 28 para. 2 and 3 (ratification of international legal instruments); Art. 29 para. 2 (financial support to political parties); Art. 33 para. 3 (issues relating to the exercise of the President's duties); Art. 48 (measures to be taken in case of a threat against national security); Art. 51 (election and composition of the Parliament); Art. 54 (electoral system) and Art. 86 (measures against serving or former members of the Cabinet or Undersecretaries for criminal offenses committed). Furthermore, the Plenum has exclusive competence over bills and proposals implementing the Constitution on the exercise and protection of individual rights, the authentic interpretation of the statutes, other matters referred to the Plenum by special provision of the Constitution or for the regulation of which a special majority is required (such as amnesty for political offences) as well as to debate and approve the budget [1: Art. 72 para. 2].

[4] Any such legislation must be carefully drafted, as established in Art. 58 et seq. entitled "Principles of Better Lawmaking [*Αρχές Καλής* Νομοθέτησης]" [3].

addition to that, they may propose the adoption of Delegated Regulations or Decrees necessary for the execution of laws in matters within their areas of competence, or enact Delegated Regulatory Acts [3: Art. 13.δ]. Consequently, for the regulation of space matters and space activity, a particularly important role is played by the competent minister, i.e. the Minister of Digital Governance, head of the Greek Ministry of Digital Governance [*Υπουργείο Ψηφιακής Διακυβέρνησης*].

At the same time, the establishment of a number of specialized directorates and/or divisions within the competent ministry are proof of the importance that the government has placed on space activity, and it goes hand in hand with the growing efforts aimed at optimizing the efficiency of the Greek space infrastructure and the related space and Earth systems.

4.1.1 A Brief History of the Ministry of Digital Governance

The Greek Ministry/minister of Digital Governance [4] is competent to propose the adoption of laws for the implementation and the enforcement of the national space policy. In this context, it is also the authority that is competent to authorize space activities, as well as to regulate the use of space resources taking into account the benefit of all persons falling within the jurisdiction of the Greek State.

More precisely, the basis for the establishment of a ministry that would be precisely responsible for the regulation and control of space-related matters appears to have been set up in the 2010s, when the Ministry of Interior and the Ministry for Infrastructure, Transport and Networks [*Υπουργείο Υποδομών, Μεταφορών και Δικτύων*] acquired the first powers on specific space matters, in addition to those of the General Secretariat for Research and Technology [*Γενική Γραμματεία έρευνας και Τεχνολογίας*].

4.1.1.1 Competent Bodies and Ministries, Under the Previous Administrations

In the early 2010s, Greece was already strongly involved in several initiatives and organizational structures for the use of space and space-based resources, mostly for the purpose of developing an information and communication network and/or to carry out remote sensing activities; e.g., the first Greek and Cypriot Telecommunications Satellite—named *Hellas Sat 2*—was launched in this context (from Cape Canaveral in 2003). However, despite these important developments, no dedicated ministry was—at that time—put in place, to be responsible for overseeing such matters.

As a result, it is the General Secretariat for Research and Technology [*Γενική Γραμματεία έρευνας και Τεχνολογίας*] which seems to have acquired *first* competences in the particular field of space, pursuant to Art. 2 of Law n° 2953/19.10.2001 [5], in a pragmatic and focused manner. Hence, in addition to being the authority that was precisely responsible for the coordination and monitoring of the implementation of the Cooperation Agreement signed between Greece and ESA on 17 January 2001 (which was ratified by means of the law n° 2953/2001), the General Secretariat for Research and Technology [*Γενική Γραμματεία έρευνας και Τεχνολογίας*] was also responsible for ensuring the payment of the annual contribution of Greece for participating in ESA programmes [5: Art. 2].[5]

Since it soon became necessary to meet the growing challenges related to the administration of space matters in a more comprehensive and global way, concrete steps were then taken in this direction and competences given to ministries. Be that as it may, in essence, the competences on space matters were initially shared among the different departments.

By way of illustration, (i) to support the increasing participation of Greece in space activities both on a practical and regulatory level, the Ministry of Interior, Decentralization and Electronic Governance [*Υπουργείο Εσωτερικών, Αποκέντρωσης και Ηλεκτρονικής Διακυβέρνησης*], as it was then known, was initially divided into the Ministry of Interior [*Υπουργείο Εσωτερικών*] and a ministry that would promote the use of space based systems for national administration, namely the Ministry of Administrative Reform and Electronic Governance, or e-Governance [*Υπουργείο Διοικητικής Μεταρρύθμισης και Ηλεκτρονικής Διακυβέρνησης*], pursuant to Presidential Decree n° 65/2011 [7].

In this context, the competences of the new Ministry of Administrative Reform and Electronic Governance were essentially the same as those exercised by the former Ministry of Interior, Decentralization and Electronic Governance, except for services that were unrelated to the powers of the new ministry. In particular, the new Ministry was now empowered to oversee the Special Department for Strategic Planning and Program Implementation [*Ειδική Υπηρεσία Στρατηγικού Σχεδιασμού και Εφαρμογής Προγραμμάτων*] and the Secretariat for Administrative Reform [*Ειδική Γραμματεία Διοικητικής Μεταρρύθμισης*], in addition to the National Printing Office [*Εθνικό Τυπογραφείο*]. In reality, the complex reform process that was launched at that time aimed at mainly organizing all the electronic interactions of the administration through Information-Communication-Technologies (ICT) with the purpose to improve public services, which was highly dependent on the utilization of space systems and infrastructure, among other things.

At the same time, (ii) law n° 4070/2012 [8] laid down, in its Art. 4, that the Ministry for Infrastructure, Transport and Networks [*Υπουργείο Υποδομών, Μεταφορών και Δικτύων*] would be the national administration responsible for the ITU procedures. In addition to that, Art. 4 para. 2 [8] established that the Minister for Infrastructure, Transport and Networks would represent the Greek State within the ITU, *inter alia*; the minister was also responsible for the entire range of issues which

---

[5]Pursuant to Art. 38 of Law no 4310/2014 titled *Research, Technological development and Innovation and other provisions*, it was laid down that the annual contributions of the Greek State for participating in the ESA programmes were to be entered in the regular budget of the General Secretariat for Research and Technology [*Γενική Γραμματεία έρευνας και Τεχνολογίας*] [6].

included the management of radio frequency spectrum and satellite orbits, as well as for the creation of a register for space radio communication items [8: Art. 4.2 para. (δ) and (ε)]. Said authority was furthermore empowered to decide, in joint decision with the Minister of Defense, on the adoption of the National Tables of Radio Spectrum Allocations [*Εθνικός Κανονισμός Κατανομής Ζωνών Συχνοτήτων (Ε.Κ.Κ.Ζ.Σ.)*], so as to allocate radiofrequencies to radio communication service providers and to determine the users [8: Art. 4.2 para. (θ)]. The Minister for Infrastructure, Transport and Networks was finally competent to promote research in space related fields—such as for the better use of Earth orbits—, in collaboration with any other co-competent minister(s) [8: Art. Art. 4.2 para. (ιη)].

It is interesting to note that law n° 4070/2012 reiterated the rule established in Art. 44.2 of the ITU Constitution, as it laid down that the orbital positions and their associated frequency allocations used by the Greek State are limited resources and must be regarded as a public good with an important social, cultural, and economic value [8: Art. 31]. In line with this reasoning, it was established that the management of allocated radio frequency bands is the State's sovereign right, which may in the aftermath grant rights to use them; i.e., according to a specific procedure which was established in the law [8: Art. 31 para. 2 to 6].

4.1.1.2 Structural Changes in the Following Years—Creation of a Dedicated Ministry

A few years later, following adoption of Presidential Decree n° 24/2015 [9], the Ministry of Administrative Reform and Electronic Governance [*Υπουργείο Διοικητικής Μεταρρύθμισης και Ηλεκτρονικής Διακυβέρνησης*], as it then existed, was (re)merged with the Ministry of Interior [*Υπουργείο Εσωτερικών*], the Ministry of Public Order and Citizen Protection [*Υπουργείο Δημόσιας Τάξης και Προστασίας του Πολίτη*] and the Ministry of Macedonia and Thrace [*Υπουργείο Μακεδονίας και Θράκης*] to form a new State authority, under the title of Ministry of Interior and Administrative Reorganization [*Υπουργείο Εσωτερικών και Διοικητικής Ανασυγκρότησης*].

Nonetheless, despite this amendment, the initial Ministry of Administrative Reform and Electronic Governance [*Υπουργείο Διοικητικής Μεταρρύθμισης και Ηλεκτρονικής Διακυβέρνησης*]—which had originally been established by Presidential Decree n° 65/2011 [7] and was later merged with three other ministries pursuant to Presidential Decree n° 24/2015 [9]—was soon revived to exist on its own, on the basis of Presidential Decree n° 123/2016 [10], under the name of Ministry of Administrative Reorganization [*Υπουργείο Διοικητικής Ανασυγκρότησης*]. Hence, although its new title was Ministry of Administrative Reorganization, it was essentially assuming the competences of the former Ministry of Administrative Reform and Electronic Governance [*Υπουργείο Διοικητικής Μεταρρύθμισης και Ηλεκτρονικής Διακυβέρνησης*] according to Art. 1.1 of said Presidential Decree [10].

Quite importantly though, the most substantial change under Presidential Decree n° 123/2016, which essentially reshaped the former Ministry of Administrative Reform and Electronic Governance into the Ministry of Administrative Reorganization [10], was that Art. 4 additionally created a completely new ministry which was in fact precisely competent for Digital Policy, Telecommunication and Information [*Υπουργείο Ψηφιακής Πολιτικής, Τηλεπικοινωνιών και Ενημέρωσης*]. Indeed, the newly created ministry was now empowered to oversee among other things the General Secretary of Digital Policy [*Γενική Γραμματεία Ψηφιακής Πολιτικής*], according to Art. 1(α) of said Presidential Decree; the General Secretary for Telecommunications and Posts [*Γενική Γραμματεία Τηλεπικοινωνιών και Ταχυδρομείων*] and a few other bodies involved in Posts and Telecommunications activities, on the basis of its Art. 1(β); as well as the General Secretary for Information and Communication [*Γενική Γραμματεία Ενημέρωσης και Επικοινωνίας*].

In addition to that, Art. 13 of law n° 4463/2017 [11]—which was adopted on 29 March 2017–, stipulated that all the competences of the Minister for Infrastructure, Transport and Networks [*Υπουργός Υποδομών, Μεταφορών και Δικτύων*] established in law n° 4070/2012 were now transferred to the above authority, namely the Minister for Digital Policy, Telecommunications and Information [*Υπουργός Ψηφιακής Πολιτικής, Τηλεπικοινωνιών και Ενημέρωσης*].[6] Still, more explicitly, Art. 13 para. 3 of the law mentioned that the above competent authority—namely the Minister for Digital Policy, Telecommunications and Information—would henceforth be competent to conceive and elaborate the national space and satellite strategy; determine all the actions and measures necessary for the development, promotion and use of satellite communication systems; or for the adoption of initiatives in these fields—including the registering of space objects in a registry, and the representation of the Greek State at the European and International level—, as well as for the coordination of operators (established under either public or private law) that are involved in space applications, technologies and services (for the steps leading to the establishment of the ministry empowered to oversee space matters, see Fig. 4.1).

4.1.1.3 Presidential Decree n° 82/2017—Defining the Ministry's Powers

The Ministry established in 2016—that is, the Ministry competent for Digital Policy, Telecommunications and Information—was given important powers, and a better-specified mandate in its key role of promoting and overseeing domestic emerging space activities. The importance that space acquired for Greece was additionally exemplified by the fact that a fully dedicated national authority for space matters was

[6] Art. 13 para. 1 of the Law no 4463/2017 precisely mentioned that wherever the term "Minister for Infrastructure, Transport and Networks [*Υπουργός Υποδομών, Μεταφορών και Δικτύων*]" is being used, in law n° 4070/2012, it shall be replaced by "Minister for Digital Policy, Telecommunication and Information matters [*Υπουργός Ψηφιακής Πολιτικής, Τηλεπικοινωνιών και Ενημέρωσης*]".

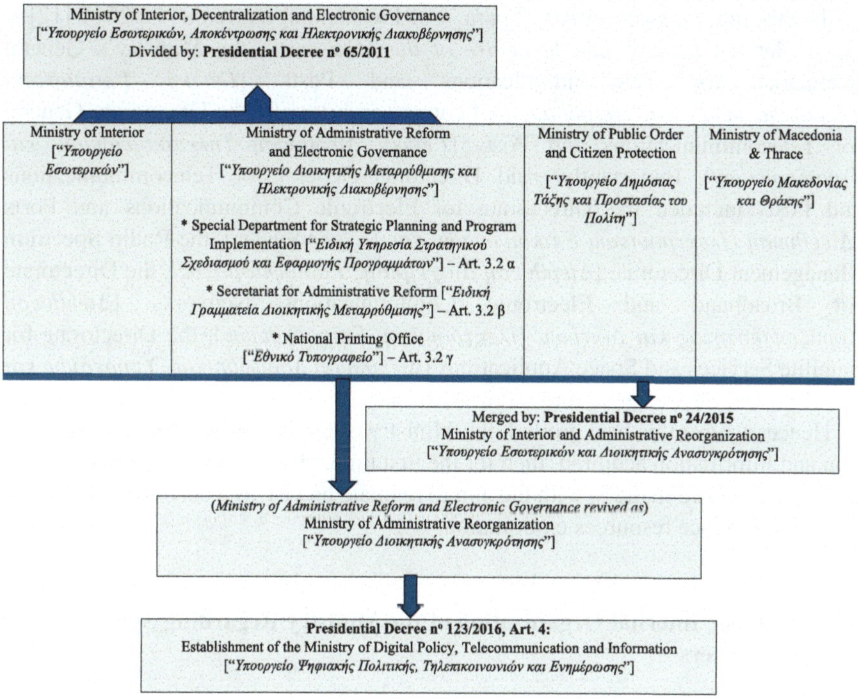

Fig. 4.1 Brief history of the Ministry of Digital Policy, Telecommunications and Information

created then, following adoption of Art. 18 of Law n° 4508/2017—under the name of Hellenic Space Agency [*Ελληνικός Διαστημικό Οργανισμός*][7]—, and was explicitly placed under the direct supervision of said ministry [10].

The specific powers of the new Ministry for Digital Policy, Telecommunications and Information—in reference to space-related matters—were established, in more detail, in Presidential Decree n° 82/2017 [12] entitled *Organization of the Ministry of Digital Policy, Telecommunication and Information*. The general competences of the Ministry were described in Article 1 of the Presidential Decree, mentioning *inter alia* that the new authority was in charge of developing and implementing a national policy and a regulatory framework (both at the European and International level), with the aim to promote the communication and digital policy, including the development of a broad spectrum of high-quality of postal services under conditions of undistorted competition [12: Art. 1.α].

In addition to that, said Ministry was entrusted with the elaboration and implementation of well-adapted policies, aimed at promoting the development of an efficient national infrastructure that would enhance the accessibility to information and remote communication services, and the processing of data transmitted using specific signals.

[7] See for more details, under Sect. 4.2.

To this end, pursuant to Art. 2 para. 4 of Presidential Decree n° 82/2017 [12]—i.e., under the general title *Structure of the Ministry*—, the Ministry's General Secretariat for Telecommunications and Posts [*Γενική Γραμματεία Τηλεπικοινωνιών και Ταχυδρομείων*] comprised *inter alia* the Directorate-General for Telecommunications and Posts [*Γενική Διεύθυνση Τηλεπικοινωνιών και Ταχυδρομείων*]. Importantly, said Directorate-General for Telecommunications and Posts included the Directorate for Electronic Communications and Posts [*Διεύθυνση Ηλεκτρονικών Επικοινωνιών και Ταχυδρομείων*], the Radio Spectrum Management Directorate [*Διεύθυνση Διαχείρισης Ραδιοφάσματος*], the Directorate for Broadband and Electronic Communications Networks [*Διεύθυνση Ευρυζωνικότητας και Δικτύων Ηλεκτρονικών Επικοινωνιών*], the Directorate for Satellite Services and Space Applications [*Διεύθυνση Δορυφορικών Υπηρεσιών και Διαστημικών Εφαρμογών*] and a few other ones.

Hence, against this background, the Ministry of Digital Policy, Telecommunication and Information acquired, then for the first time, broad and well-defined powers over space-related matters, with the aim to regulate precise aspects of national space activity and space resources exploitation.

4.1.1.4 Basic Internal Organization of the Ministry Regarding Space Matters

Fundamentally, the General Secretariat for Telecommunications and Posts [*Γενική Γραμματεία Τηλεπικοινωνιών και Ταχυδρομείων*] became one of the key services responsible for major space matters within the Ministry of Digital Policy, Telecommunication and Information [*Υπουργείο Ψηφιακής Πολιτικής, Τηλεπικοινωνιών κι Ενημέρωσης*]. In particular, said General Secretariat was competent for overseeing many of the activities of the Directorate-General for Telecommunications and Posts [*Γενική Διεύθυνση Τηλεπικοινωνιών και Ταχυδρομείων*], while it was also empowered to initiate action such as to specifically address matters related to space activity.

Precise rules on the organization and functioning of the General Directorates of the Ministry of Digital Policy, Telecommunication and Information (including on their own directorates) were laid down in the Presidential Decree n° 82/2017 [12]. In this context, Chapter IV [*Κεφάλαιο Δ'*] of said Presidential Decree, namely Art. 16 to 23, regulated in detail the organization and coordination of the Directorate-General for Telecommunications and Posts, as well as of its own Directorates.

By way of illustration, Art. 17 stipulated—as regards the powers of the Directorate-General for Telecommunications and Posts [*Γενική Διεύθυνση Τηλεπικοινωνιών και Ταχυδρομείων*]—that this service would specifically focus on ensuring the optimal use of radiofrequencies, on developing wireless telecommunication services, satellites and broadcasting services [12: Art. 17.1.δ]. Towards this objective, the Directorate-General for Telecommunications and Posts primarily comprised four directorates specifically responsible for space matters, and outlined

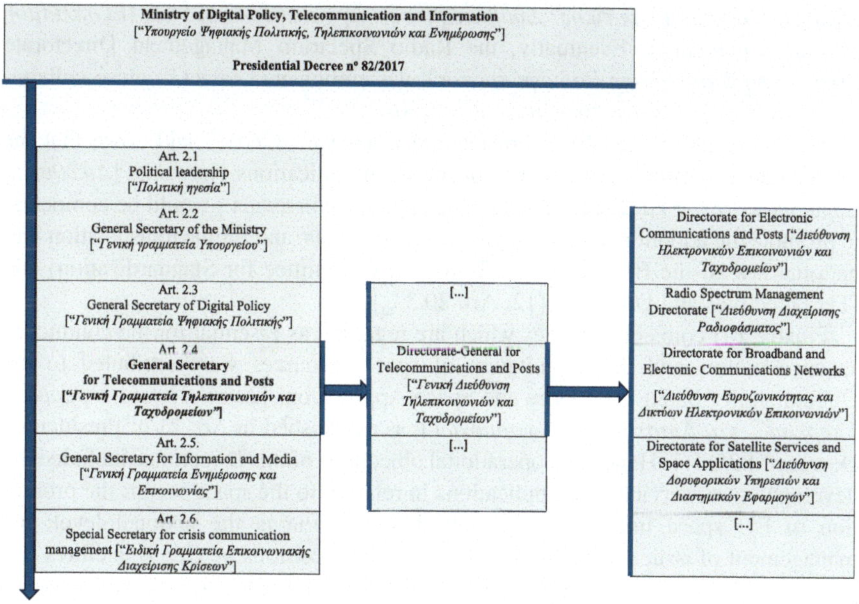

Fig. 4.2 Structure of the Directorate-General for Telecommunications and Posts, of the Ministry of Digital Policy, Telecommunications and Posts

below (for the ministry's organization chart, detailing services competent in space matters, see Fig. 4.2).

(1) Regarding the exploitation of space resources and systems, Art. 18 mentioned that the Directorate for Electronic Communications and Posts [*Διεύθυνση Ηλεκτρονικών Επικοινωνιών και Ταχυδρομείων*] would be competent to monitor the short-term changes on merchant ships and Earth costal stations radiocommunication, as well as to inform the International Telecommunication Union (ITU) on such matters [12: Art. 18.5.ι].

(2) Against this background, Art. 19 para. 2 of the Presidential Decree n° 82/2017 laid down that a particular body, the Radio Spectrum Management Directorate [*Διεύθυνση Διαχείρισης Ραδιοφάσματος*], would be entitled to promote the protection and optimal use of radiofrequencies, such as by ensuring that the ITU Radio Regulations were being applied [12: Art. 19.3.β]; to secure the international coordination procedure of the radio spectrum and to conclude agreements aimed at the suppression of harmful radio interference [12: Art. 19.3.γ]; to notify the ITU about any Radio Frequency Assignments [12: Art. 19.3.ε], to elaborate and adopt the National Board for Radio Frequency Assignments [12: Art. 19.3.στ], etc. Practically, this Directorate was composed of the Division for Fixed and Mobile services Spectrum Management [*Τμήμα Διαχείρισης Φάσματος Σταθερών και Κινητών Υπηρεσιών*], the Radio and Television Spectrum Management Division [*Τμήμα Διαχείρισης Φάσματος Ραδιοφωνίας και Τηλεόρασης*] and the Division for Electromagnetic Compatibility, Infrastructure and Radio Communications Equipment

[*Τμήμα Ηλεκτρομαγνητικής Συμβατότητας, Εγκαταστάσεων και Εξοπλισμού Ραδιοεπικοινωνιών*]. Eventually, the Radio Spectrum Management Directorate [*Διεύθυνση Διαχείρισης Ραδιοφάσματος*] was empowered to act in close collaboration with its Divisions, to achieve its purpose.

(3) Furthermore, Art. 20 of Presidential Decree n° 82/2017 laid down that the Directorate for Broadband and Electronic Communications Networks [*Διεύθυνση Ευρυζωνικότητας και Δικτύων Ηλεκτρονικών Επικοινωνιών*] would be competent to promote the adoption of laws and implementing acts, and for the transposition into national law of the EU, CEN (i.e., European Committee for Standardization) and ITU Directives and Decisions [12: Art. 20.3.ζ.].

Finally, (4) core competences which are regarded as essential for the conduct of space activities and the exploitation of space resources were attributed to the Directorate for Satellite Services and Space Applications [*Διεύθυνση Δορυφορικών Υπηρεσιών και Διαστημικών Εφαρμογών*], as established in Art. 21 of Presidential Decree n° 82/2017. Hence, the operational objectives of this Directorate included the development of services and applications in relation to the space sector; the promotion of EU space initiatives and their dissemination at the national level; the management of issues related to the use of orbital positions by Greek satellites and the provision of public satellite services [12: Art. 21.1.α to 21.1.δ]. To be operational, the Directorate for Satellite Services and Space Applications comprised (i) the Satellite Communication Division [*Τμήμα Δορυφορικών Επικοινωνιών*]; (ii) the Division for Public Satellite Services [*Τμήμα Παροχής Δορυφορικής Κυβερνητικής Υπηρεσίας*] and (iii) a Division for the Management of Users' Rights to Public Satellite Services [*Τμήμα Διαχείρισης Δικαιωμάτων Πρόσβασης σε Δορυφορική Κυβερνητική Υπηρεσία*], [12: Art. 21.2] — all of them are further detailed hereafter.

Specifically, (i) the Satellite Communication Division was *first* competent to promote the elaboration and updating of the legal framework on satellite communication and space policy, in close collaboration with its own Divisions and/or with other competent authorities [12: Art. 21.3.α]. It was additionally competent to notify the International Telecommunication Union (ITU) regarding any requests to use orbital positions and radio frequencies, as well as for the management of any orbital positions and radio frequencies issues [12: Art. 21.3.β]; for the planning of policies on the use of the radio spectrum for satellite communication services, and for the development of a broader space policy framework, more particularly by collecting and assessing the needs of public bodies and authorities [12: Art. 21.3.γ]. Lastly, it was empowered to decide on the creation and management of a Record/Registry for Space Stations and Space Objects, the management of issues relating to space activity and space objects and a Record of all the notifications sent to the ITU (i.e., ITU filings) [12: 21.3.ι].

By way of comparison, (ii) the Division for Public Satellite Services [*Τμήμα παροχής δορυφορικής κυβερνητικής υπηρεσίας*] was now competent to deal with such practical matters as cooperation with national and European bodies regarding the management of users' issues [12: Art. 21.4. α]; the management and assessment of users' safety issues [12: Art. 21.4. β]; the management of harmful electromagnetic interferences on the Public Regulated Service/PRS [12: Art. 21.4.γ] etc.

Finally, the last division, namely (iii) the Management of Users' Right to Public Satellite Services Division [*Τμήμα διαχείρισης δικαιωμάτων πρόσβασης σε δορυφορική κυβερνητική υπηρεσία*] was empowered to determine the users' conditions of access to the Public Regulated Service/PRS, either for individuals or groups of individuals [12: Art. 21.5.α]. Furthermore, it was competent to also submit proposals aimed at developing synergies in the field of telecommunications networks and/or to establish with any other European and/or international Space Agencies, collaborations *inter alia* in the fields of transportation, borders control, naval operations and security [12: Art. 21.5.ζ]; as well as to participate in, and benefit from, collaboration in the fields of space activity and satellite communication, and to integrate into national infrastructure the EU space policy, in collaboration with other competent authorities [12: Art. 21.5.η].

4.1.2 Creation and Competences of the Ministry of Digital Governance

In 2019, the Greek State went one step further. Art. 1 of the Presidential Decree n° 81/2019 [13] abolished the then-existing Ministry of Digital Policy, Telecommunications and Information [*Υπουργείο Ψηφιακής Πολιτικής, Τηλεπικοινωνιών και Ενημέρωσης*]. It established, in its place, the Ministry of Digital Governance [*Υπουργείο Ψηφιακής Διακυβέρνησης*] with an enhanced political role, as the new entity was specifically in charge of monitoring key space matters, among other things.

In fulfillment of its mission—described in the 2019 Presidential Decree—, the newly created ministry absorbed quite a few services of the former Ministry of Digital Policy, Telecommunications and Information, that were competent for space matters (and had gained relevant expertise); such as, mainly, the General Secretariat of Digital Policy [*Γενική Γραμματεία Ψηφιακής Πολιτικής*] and the General Secretariat for Telecommunications and Posts [*Γενική Γραμματεία Τηλεπικοινωνιών και Ταχυδρομείων*].

As a result, the new Ministry for Digital Governance had now authority over the main bodies which were until then entitled to regulate, plan, and implement space law and space policy matters, as laid down in the Presidential Decree n° 82/2017, with all their Directorates and competences [13: Art. 1.1.2].[8] In addition to that, the

[8]Such as the General Secretary for Telecommunications and Posts [*Γενική Γραμματεία Τηλεπικοινωνιών και Ταχυδρομείων*] which included the Directorate General for Telecommunications and Posts [*Γενική Διεύθυνση Τηλεπικοινωνιών και Ταχυδρομείων*], composed by the Directorate for Electronic Communications and Posts [*Διεύθυνση Ηλεκτρονικών Επικοινωνιών και Ταχυδρομείων*], the Radio Spectrum Management Directorate [*Διεύθυνση Διαχείρισης Ραδιοφάσματος*], the Directorate for Broadband and Electronic Communications Networks [*Διεύθυνση Ευρυζωνικότητας και Δικτύων Ηλεκτρονικών Επικοινωνιών*] and the Directorate for Satellite Services and Space Applications [*Διεύθυνση Δορυφορικών Υπηρεσιών και Διαστημικών Εφαρμογών*].

Ministry of Digital Governance acquired the supervision of the national space agency, known then as the Hellenic Space Agency [*Ελληνικός Διαστημικό Οργανισμός*], [13: Art. 1.2.1 (γ)], which was subsequently reshaped into the Hellenic Space Center [*Ελληνικό Κέντρο Διαστήματος*], according to Art. 60 para. 7 of Law 4623/09.08.2019 [14].

Following that, the organization and functioning of the new ministry was clarified by means of the Presidential Decree n° 40/2020 [15] on the Organization of the Ministry of Digital Governance. The new authority comprises, now, according to Art. 2.1.στ. of the 2020 Presidential Decree, several of the previous services specialized in space matters which, however, were reorganized differently this time. By way of illustration, the General Secretary for Telecommunications and Posts [*Γενική Γραμματεία Τηλεπικοινωνιών και Ταχυδρομείων*] is under this new scheme, in charge of fewer but highly specialized services, which principally include the Directorate-General for Telecommunications and Posts [*Γενική Διεύθυνση Τηλεπικοινωνιών και Ταχυδρομείων*] and a Directorate-General for cyber-security [*Γενική Διεύθυνση Κυβερνοασφάλειας*], in addition to the Independent Division for Internal Coordination and Public Relations [*Αυτοτελές Τμήμα Εσωτερικού Συντονισμού και Διεθνών Σχέσεων*].

The principal competences of the Directorate-General for Telecommunications and Posts [*Γενική Διεύθυνση Τηλεπικοινωνιών και Ταχυδρομείων*] and of its Divisions—in the particular areas of space and space-related resources utilization—were laid down in further detail in Chap. 6 [*Κεφάλαιο ΣΤ'*] Art. 37 et seq., and they are succinctly presented below.

4.1.2.1 The Directorate-General for Telecommunications and Posts, a Key Service for Space Matters

According to Art. 38 of Presidential Decree n° 40/2020, the Directorate-General for Telecommunications and Posts [*Γενική Διεύθυνση Τηλεπικοινωνιών και Ταχυδρομείων*]—i.e., which has as its principal objective to develop reliable, safe, and competitive telecommunication and postal services—, is empowered to mainly ensure the optimal use of the radio spectrum and satellite services [15: Art. 38.1 (δ).], to represent the country and to as well coordinate representatives of co-competent authorities in all the relevant European and International bodies [15: Art. 38.1 (ζ)].

Regarding its organization, Art. 38.2 also clarified that the Directorate-General for Telecommunications and Posts consists of the Directorate for Wireless Communication Services, 5G Networks and Space [*Διεύθυνση Ασύρματων Επικοινωνιών, Δικτύων Πέμπτης Γενιάς (5G) και Διαστήματος*], the Directorate for Broadband and Electronic Communications Networks [*Διεύθυνση Ευρυζωνικότητας και Δικτύων Ηλεκτρονικών Επικοινωνιών*], the Directorate for Technological Equipment and Infrastructure [*Διεύθυνση Τεχνολογικού Εξοπλισμού και Εγκαταστάσεων*] and the Posts Division [*Διεύθυνση Ταχυδρομείων*].

In this context, it appears that the competent authority to regulate issues arising from the commercial exploitation of space resources is mainly the Directorate for

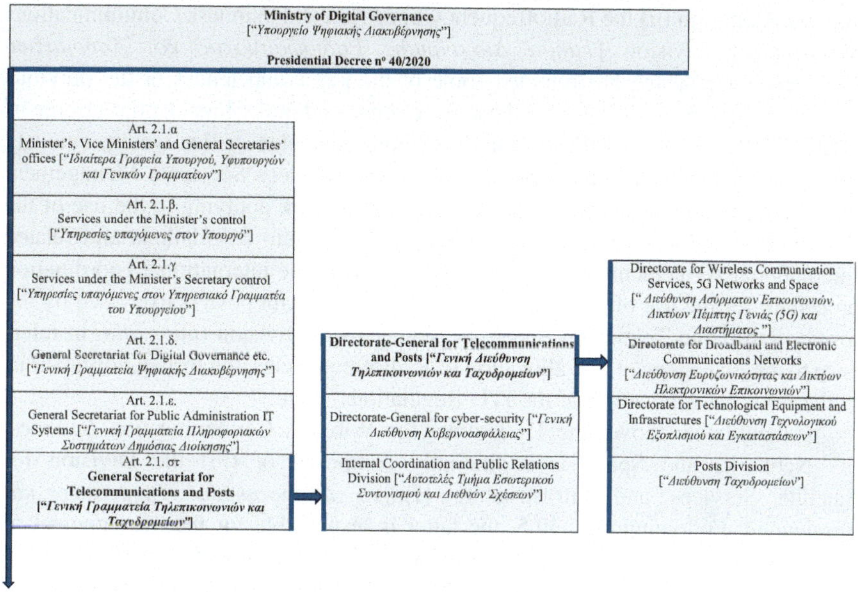

Fig. 4.3 Structure of the Ministry of Digital Governance

Wireless Communication Services, 5G Networks and Space. Hence, Art. 39 precisely mentions that it contributes to the objectives of ensuring optimal use of radio frequencies, laying down the basis for radio spectrum policy; managing issues related to the use of orbital positions and encouraging measures aimed at the development and use of the national and European space infrastructure [15: Art. 39.1].

From an organizational perspective, in order to achieve its aims, said Directorate is composed of (i) the 5G Wireless Communication Services Division [*Τμήμα Ασύρματων Επικοινωνιών Πέμπτης Γενιάς (5G)*], (ii) the Radiofrequency Spectrum and Wireless Communications Management Division [*Τμήμα Διαχείρισης Ραδιοφάσματος και Ασύρματων Επικοινωνιών*] and (iii) the Division for Satellite Services and Infrastructure [*Τμήμα Δορυφορικών Υπηρεσιών και Υποδομών*] (for the current structure of the Directorate-General for Telecommunications and Posts, see Fig. 4.3).

Regarding their specific means of action, (i) the 5G Wireless Communication Services Division [*Τμήμα Ασύρματων Επικοινωνιών Πέμπτης Γενιάς (5G)*] primarily focuses on the development and management of 5G systems, infrastructures and on fostering the fulfillment of all the related rules and procedures.[9] As a synergy of actions must be taken to reach these objectives, said Division may act in

[9]For example, it is competent to follow the developments in relation to 5G technology and to prepare a national program allowing the integration and use of 5G systems [15: Art. 39.3].

collaboration with (ii) the Radiofrequency Spectrum and Wireless Communications Management Division [*Τμήμα Διαχείρισης Ραδιοφάσματος και Ασύρματων Επικοινωνιών*], which has retained some of the key competences of the previous competent bodies in this field, as these were organized under Presidential Decree n° 82/2017 [12]. By way of illustration, it assumes competences that were previously exercised by the Division for Fixed and Mobile services Spectrum Management (e.g., such as to process and update the legal framework applying to the use of the Radiofrequency Spectrum; to ensure that the ITU Radio Regulations and related International Agreements are duly applied; to secure the international coordination procedure for the use of the radio spectrum etc.[10]); certain powers exercised by the former Radio and Television Spectrum Management Division (*inter alia*, in reference to the management and allocation of Radiofrequencies, the formal notifications to the ITU, the application of the ITU Regulations[11]) etc.

As mentioned above, the Directorate for Wireless Communication Services, 5G Networks and Space is additionally composed of (iii) the Division for Satellite Services and Infrastructure [*Τμήμα Δορυφορικών Υπηρεσιών και Υποδομών*]. Following Art. 39.5, the latter is responsible for the management of space-related matters, assuming again most of the competences that were exercised by the previously competent body (i.e., by the Directorate for Satellite Services and Space Applications, as it was organized by Art. 21.3 of Presidential Decree n° 82/2017 [12]).

Therefore, the new Division for Satellite Services and Infrastructure is competent to process and update the legal framework regulating satellite communication and space policy, together with other competent authorities [15: Art. 39.5 (α)]; to notify the ITU on requests to use orbital positions and radio frequencies, as well as to decide on orbital positions and radio frequencies issues [15: Art. 39.5 (β)]; to oversee the coordination of conditions for the availability and efficient use of radiofrequencies [15: Art. 39.5 (στ)]; or to establish, keep up to date and manage a Record/Registry for Space Stations and Space Objects, in addition to deciding on issues relating to space activity and space objects [15: Art. 39.5 (η)]. The Division is likewise empowered to represent the country in EU and other bodies in reference to space and space-related matters [15: Art. 39.5 (θ)].

4.1.2.2 The Particular Competences of the Directorate-General for Cyber-Security

It should be noted that, with regard to the identification of competent authorities in the field of space matters *lato sensu*, an equally important role is played

[10] Art. 39.4 of the Presidential Decree 40/2020 is reiterating some of the provisions of Art. 19.3 of the Presidential Decree 82/2017.

[11] Art 39.4 of the Presidential Decree 40/2020 is also reiterating many of the provisions of Art. 19.4 of the Presidential Decree 82/2017.

by the Directorate-General for cyber-security [*Γενική Διεύθυνση Κυβερνοασφάλειας*]. Essentially, Art. 43.2 of Presidential Decree 40/2020 [15] stipulates that said Directorate-General consists of the Directorate for the Strategic Planning of Cyber-security [*Διεύθυνση Στρατηγικού Σχεδιασμού Κυβερνοασφάλειας*], the Directorate for Prevention and Protection [*Διεύθυνση Πρόληψης και Προστασίας*], the Directorate for Business Continuity [*Διεύθυνση Επιχειρησιακής Συνέχειας*] and finally the Directorate for Coordination between Operators [*Διεύθυνση Συντονισμού Φορέων*]. To a certain extent, the titles of these Directorates give more or less an idea of their purposes and competences.

In this context, it is noteworthy that the Directorate for the Strategic Planning of Cyber-security is composed, *inter alia*, by the Public Satellite Services Division [*Τμήμα Κυβερνητικής Δορυφορικής Υπηρεσίας*] which is explicitly responsible for the safe management of users' and public administration material, and the safe connection with the Galileo Security Monitoring Center [15: Art. 44.6 (α)]; the management of any harmful interference with public satellite services [15: Art. 44.6 (γ)] and for all the access conditions to these services.

4.1.3 Further Transfer of Competences to the Ministry—Law no 4727/2020

Building on these initial steps, law no 4727/2020 on Digital Governance matters [16] focused even more intently on organizing the safe and efficient access to digital products and related services, within the wider context of public administration.[12] As a result, the Ministry of Digital Governance was reaffirmed as having the principal role in drawing the policy and coordinating efforts in relation to the use of radio frequencies and the regulation of telecommunications—such as, with the aim of harmonizing the use of radio spectrum—, in collaboration with EU member States and EU institutions and authorities [16: Art. 112.1 to 112.3; Art. 114.2 etc.].[13] In addition to that, the Ministry of Digital Governance was also established as the correspondent Greek Administration for telecommunications [*Αρχή Τηλεπικοινωνιών*], as provided for in the ITU Constitution.

In exercising its new competences under law no 4727/2020, the Ministry of Digital Governance may as well obtain assistance from a separate body (partially competent for space-related matters), that is the Hellenic Telecommunications &

[12] Amongst other topics, the Law n° 4727/2020 regulates the use of electronic signatures and electronic time stamps; the collection and processing of private data; digital public service and infrastructure; accessibility requirements, personal identification numbers and electronic documents matters; the statistical processing of data and electronic trust services; open data and the reuse of public sector information etc.

[13] As regards the establishment and functioning of the internal market in electronic communications, the Ministry of Digital Governance cooperates, where needed, with the Hellenic Telecommunications & Post Commission [*Εθνική Επιτροπή Τηλεπικοινωνιών και Ταχυδρομείων*].

Post Commission [*Εθνική Επιτροπή Τηλεπικοινωνιών και Ταχυδρομείων*]. More precisely, the latter—namely the Hellenic Telecommunications & Post Commission—was established by law no 2246/1994, as amended by law no 4070/2012, as the National Regulatory Authority [*Εθνική Ρυθμιστική Αρχή*] for the provision of networks, infrastructure and electronic communication services, while overseeing the radio frequency spectrum use management [16: Art. 115.9].[14]

As a result, the Hellenic Telecommunications & Post Commission is the competent authority for—among other things—keeping and updating the National Registry of Radiofrequencies [*Εθνικό Μητρώο Ραδιοσυχνοτήτων*] which is, as clarified in Art. 113.2 (ιδ) and Art. 114 of law no 4727/2020, the database containing all the radio spectrum ranges allocated at the national level [16]. The requirements and rules on the keeping of the National Registry of Radiofrequencies [*Εθνικό Μητρώο Ραδιοσυχνοτήτων*] aim at allowing the Registry to fulfill its function (as it is defined in said Art. 114); such rules are laid down by a joint decision of the Minister of Digital Governance and the Minister of Defense, upon advice of the Hellenic Telecommunications & Post Commission [16: Art. 113.2 (ιδ).]. In addition to that, said authority is empowered to define the radiofrequencies or radio frequency bands requiring a right to use, in compliance with law no 4727/2020 [16: Art. 113.2 (ιζ)]; to grant, revoke or limit rights of use of radiofrequencies for the provision of networks and electronic communication services, as well as for the use of radio equipment, in accordance with Art. 113.2 (ιη) of law no 4727/2020; and to manage the radiofrequency spectrum or radio frequency bands for which it is responsible, on the basis of the applicable legislation [16: Art. 113.2 (ιθ)].

Be that as it may, it is the competent ministry/minister which plays the central role in handling legal and practical challenges in the realm of space activities, while remaining one of the most influential authorities in the context of space-related policymaking. Against this background, the minister's principal task is to also coordinate and give effective direction to the several services and directorates of the ministry that are responsible for law enforcement and policymaking, as briefly presented below.

4.1.3.1 Specific Competences of the Minister of Digital Governance

The Minister of Digital Governance was vested—by law no 4727/2020—with notably broad powers, taking into account the particularities and key features of space activity. For instance, Art. 114 provides him/her with substantial competences in this domain, such as to elaborate the electronic communications policy [16: Art. 114.2 (α)]; to assess the necessity of carrying out further investments in this field, in collaboration with other co-competent ministers [16: Art. 114.2 (γ)]; to represent the

[14] Taking also into account the precise content of the National Tables of Radio Spectrum Allocations [*Εθνικός Κανονισμός Κατανομής Ζωνών Συχνοτήτων (Ε.Κ.Κ.Ζ.Σ.)*] and the rules laid down in Art. 113 and 114 [16: Art. 115.9].

country in the ITU (together with other national authorities or co-competent ministries, where relevant); to transpose the ITU Recommendations and relevant—EU and other—legal instruments into national law, by adopting ministerial decisions [16: Art. 114.2 (δ) and (ε)]; to address the issues arising from the use of orbital positions which have been allocated to Greece[15] and to promote the harmonization of the use of radio frequencies across the European Union, in particular with the aim to ensure their efficient utilization.[16]

In addition to that, the Minister of Digital Governance is entitled to decide on the allocation and sharing of the radio frequency spectrum, at the national level, together with the Minister of Defense. The two of them shall adopt, following a joint decision, the National Tables of Radio Spectrum Allocations [*Εθνικός Κανονισμός Κατανομής Ζωνών Συχνοτήτων (Ε.Κ.Κ.Ζ.Σ.)*]; thereby allocating radiofrequencies to radio communication service providers, and determining users—following a case-by-case approach—as well as the administrative and technical conditions of such use [16: Art. 114.2 para. θ]. The specific content and requirements of the National Tables of Radio Spectrum Allocations are precisely laid down in Art. 115.6. Still, the National Tables of Radio Spectrum Allocations may be amended—again by a joint decision of the Minister of Digital Governance and the Minister of Defense—when necessary, in cooperation with the National Commission for Telecommunications and Posts [*Εθνική Επιτροπή Τηλεπικοινωνιών και Ταχυδρομείων*], [16: Art. 115.7].

More importantly, the Minister of Digital Governance is in charge of policymaking in reference to the national space and satellite policy and strategy. He/She is empowered to adopt measures concerning the development, promotion, and exploitation of satellite communication systems, as well as to plan and carry out relevant activities in this field, in cooperation with the Minister of Defense [16: Art. 114.2 (κα)]. Furthermore, the Minister of Digital Governance is competent to decide on the registration of space objects in the relevant registry; on the representation of the country for space matters, at the European and international level; and on the coordination (where necessary) of the public and private stakeholders involved in space matters, such as in the fields of space applications, technology, and services [16: Art. 114.2 (κα)].

Finally, the Minister of Digital Governance is the competent authority for granting rights to use the radio spectrum, and for deciding on conditions and procedures required for such authorization and use [16: Art. 114.2 (κβ)]. He/She is also entrusted with ensuring the payment of the Greek State's participation in the ESA mandatory activities and optional programs. Importantly, as regards said participation, Art. 119.9 lays down that its purpose must be to strengthen the national

[15] These competences appear to complement the powers of the Division for Satellite Services and Infrastructure, as established by the Presidential Decree n° 40/2020 [15: Art. 39.5 (β)].

[16] These competences as well, seem to supplement the responsibilities of the Division for Satellite Services and Infrastructure, as established by the Presidential Decree n° 40/2020 [15: Art. 39.5 (στ)].

defense and security [16: Art. 119.9 point α]; to promote the development of the Greek space industry [16: Art. 119.9 point β],[17] to support the development of space data and applications [16: Art. 119.9 point γ] and to promote research and innovation in these fields [16: Art. 119.9 point δ].

In exercising all the powers mentioned above, the Ministry of Digital Governance (i.e., and the minister) is supported by the General Secretariat for Telecommunications and Posts [*Γενική Γραμματεία Τηλεπικοινωνιών και Ταχυδρομείων*].[18] Moreover, Art. 119.9 para Δ. establishes that the ministry may set up a Seven-Member Committee competent to give advice in reference to the participation of Greek operators in ESA activities, programs, and on any other related issue [16: Art. 119.9 para. Δ], as well as to inform the Minister of Digital Governance on the evaluation of Greek operators' proposals, based on the results published by the ESA.[19]

Hence, against the background of this institutional framework which is steadily being developed since the adoption of the first laws and presidential decrees (e.g., in 2017 and 2019)—establishing the general competence of said authority to decide on space matters, taking into account the interests of the Greek State (as well as of domestic stakeholders) while promoting and fostering international collaborations in this domain—, the competent minister took *inter alia* the initiative of signing frameworks of agreement with several States. In particular, a number of joint programs and MoUs of agreement (setting short-term and medium-term goals) were agreed upon and they aimed, essentially, at creating viable as much as promising synergies with key foreign space agencies.

To only provide one among many examples of such collaboration agreements, on 9 April 2019 a Joint Statement on collaboration was signed, between the Hellenic Space Agency S.A. [*Ελληνικός Διαστημικός Οργανισμός Α.Ε.*] and the US National Aeronautics and Space Administration (NASA), on the occasion of the Global Space Symposium. More precisely, the agreement mentioned the possibility for Greece of taking part in activities organized under the NASA program *Advanced Human and Robotic Exploration ON and AROUND the MOON*, and to facilitate cooperation

[17] The conditions for granting letters of support to Greek Space operators concerned by ESA's activities and programs are also laid down in the law [16: Art. 119.9 para. B].

[18] The General Secretary for Telecommunications and Posts [*Γενική Γραμματεία Τηλεπικοινωνιών και Ταχυδρομείων*] is competent to undertake the planning, representation, management and coordination of actions and measures related to the participation of Greece in ESA activities and programs. Within this purpose, the General Secretariat for Telecommunications and Posts cooperates with the ESA in matters related to the Greek participation in ESA's programs, as well as on the evaluation and support of Greek operators' participation [16: Art. 119.9 para. A].

[19] This is deemed necessary, in order for the Minister of Digital Governance to be able to issue a letter of support, in accordance with Law no 4727/2020, Article 119.9 para Γ. The Seven-Member Committee of Article 119.9 para Δ. may decide differently than the ESA, and meets as provided for in Article 119.9 para Δ (β).

between the two States in all fields of space activity [17].[20] In this context the launch of the first Greek robotic spacecraft, including sending it to the Moon, was foreseen for 2022. The Greek participation was entitled *Hellas to the Moon*, and its purpose was to explore the lunar surface to collect data that could be used by universities and researchers [18].

4.1.3.2 The Ministry's Explicit Powers: Use of Allocated Radio Frequency Spectrum

Law no 4727/2020 conferred, in parallel to the above powers attributed to the minister, explicit competence to the Ministry of Digital Governance in the field of management and use of the radio frequency spectrum. Within this specific objective, the law highlights, initially—in its Art. 115.1—the fact that the radio spectrum (similarly to the orbital positions, both GSO and non-GSO) must be considered as a limited natural resource, as established in Art. 44.2 of the ITU Constitution [19].

Indeed, Art. 44.2 of the ITU Constitution, reads that: "In using frequency bands for radio services, Member States shall bear in mind that radio frequencies and any associated orbits, including the geostationary-satellite orbit, *are limited natural resources* and that they must be used rationally, efficiently and economically, in conformity with the provisions of the Radio Regulations, so that countries or groups of countries may have equitable access to those orbits and frequencies, taking into account the special needs of the developing countries and the geographical situation of particular countries" (emphasis added).

In the same vein, Art. 115 of law no 4727/2020 mentions that the radio spectrum is a limited resource (while the management of the allocated radio frequency bands is the State's sovereign right) and a public good that has an important social, cultural, and economic value [16: Art. 115.1]. In addition to that, it clarifies that the radio frequency spectrum must be essentially used for very specific purposes, such as principally to safeguard human life and property [16: Art. 115.2 (α)]; to serve national security and defense [16: Art. 115.2 (β)]; and to provide support for combating crime and maintaining the rule of law [16: Art. 115.2 (γ)]. Furthermore, it may be used for social and economic development—by providing efficient communication services, both at the national and international level, either for personal or professional use [16: Art. 115.2 (δ)]—as well as to facilitate the sustainable management of natural resources [16: Art. 115.2 (ζ)].

The approaches to management and supervision—of the uses of the radio frequency spectrum—are detailed in Art. 115.3: such control must be conducted by the

[20] More precisely, "Under the deal, Greece [*would*] also take part in the Lunar Orbital Platform-Gateway (LOP-G), an international, NASA-led project to create a lunar-orbit space station in the 2020s. It also foresees exchanges of information and data for academic and scientific research, as well as the participation of Greek educational institutions in international programs, among other benefits" [17].

competent bodies and authorities, in line with national requirements, international and ITU agreements, while considering EU laws.[21] Interestingly, Art. 115.5 reaffirms that, in applying the law, account must be taken of *all* the international agreements (such as the ITU Radio Regulations) and objectives of public security. To achieve this goal, the authorities which are entitled to ensure the efficient management of the radio spectrum, in line with Art. 111 to 114 [16: Art. 115.4], are the Ministry of Digital Governance and the Hellenic Telecommunications & Post Commission [*Εθνική Επιτροπή Τηλεπικοινωνιών και Ταχυδρομείων*].

In this context (as mentioned above under Sect. 4.1.3.1), the Minister of Digital Governance adopts, by a joint decision with the Minister of Defense, the National Tables of Radio Spectrum Allocations [*Εθνικός Κανονισμός Κατανομής Ζωνών Συχνοτήτων (Ε.Κ.Κ.Ζ.Σ.)*]. Following on from this, the Ministry of Digital Governance provides on its official website the detailed content of the National Tables of Radio Spectrum Allocations [*Εθνικός Κανονισμός Κατανομής Ζωνών Συχνοτήτων (Ε.Κ.Κ.Ζ.Σ.)*], and informs the European Agency for Communication [*Ευρωπαϊκό Γραφείο Επικοινωνιών*] and the European Commission, accordingly.[22] At the same time, the Hellenic Telecommunications & Post Commission [*Εθνική Επιτροπή Τηλεπικοινωνιών και Ταχυδρομείων*] issues the Regulation on Conditions of use of the Radio Spectrum [*Κανονισμός Όρων Χρήσης του Ραδιοφάσματος*], to allow management of the use of the radio frequency spectrum (i.e., the precise technical and administrative conditions relating to its use, as they are defined in Art. 115.10 of law n° 4727/2020, are established there).

Hence, the Ministry of Digital Governance, together with the Hellenic Telecommunications & Post Commission, is responsible for the overall efficient use of the radio frequency spectrum by all the stakeholders and operators [16: Art. 115.11]; as well as for monitoring the use of the spectrum and for keeping informed the ITU about surveillance matters, as it is established in Art. 116 of law no 4727/2020 [16: Art. 116.2 et seq.]. Within that context, however, deciding who has the right to use orbital positions and associated frequencies, is left to the Minister of Digital Governance who decides in the last instance.

4.1.3.3 Deciding Upon the Right to Use Orbital positions

Along the same lines as Art. 115 of law no 4727/2020 (regulating the use of the radio frequency spectrum), Art. 119.1 provides that the GSO and non-GSO orbital positions—similarly to the frequency rights associated to orbital positions and allocated to the country—are a limited resource and a public good that has an

[21] The use of the radio frequency spectrum in Greece is to be organized in such a way that no other EU member State is prevented from using it [16: Art. 136].

[22] The Hellenic Telecommunications & Post Commission [*Εθνική Επιτροπή Τηλεπικοινωνιών και Ταχυδρομείων*] must also make available, on its official website, information related to rights, conditions, procedures etc. related to the use of radio frequencies [16: Art. 115.8].

important social, cultural, and economic value. Their management is recognized as being the sovereign right of the State (and must—in this sense—be dealt with at the national level). Be that as it may, in granting rights of use for orbital positions (and associated frequencies) allocated to the country by the ITU, the State is bound to issue an open and non-discriminating call for tender procedure.[23] In the event that no tender is submitted, or if the procedure is declared to be unsuccessful, the Minister of Digital Governance is entitled to initiate a negotiation procedure to grant rights of usage by direct agreement.[24]

No right of use for orbital positions (and associated frequencies) can be transferred, without prior approval from the Minister of Digital Governance [16: Art. 119.3]. In addition to that, the Minister may amend any permit or right of usage on orbital positions (and of associated frequencies), to be in line with international coordination actions [16: Art. 119.4].[25] In any case, the provision of satellite-based services using satellite systems may only be conducted in accordance with the ITU Radio Regulations and the ITU Constitution [16: Art. 119.6].

Art. 119.8 of law no 4727/2020 further empowers the Minister of Digital Governance to ensure the protection of classified information provided in this context—data, accompanying documents and other instruments related to space matters, operations and activity (*inter alia* memoranda, agreements etc.)—from public disclosure, for any reason of national security. In this event, such information will not be published in the Greek Government Gazette or made available by any means on the Internet; it will only be accessible on a case-by-case basis, under strict conditions and exclusively by persons establishing a particular interest therein [16: Art. 119.8].

Importantly, a few additional specific conditions may be attached to the granting of rights of use for any orbital positions and associated frequencies, with the purpose of ensuring the efficient utilization of these resources [16: Art. 121.1; and Art. 155 et seq.].[26] In essence, spectrum allocation used for electronic communication services and rights of use of such frequencies may only be based on objective, transparent, non-discriminatory, and proportionate criteria, while allowing to promote competition [16: Art. 153.1]. Therefore, in adopting such decisions, the competent bodies must take into consideration all relevant ITU—as well as other international law—obligations and commitments, and any public security objectives likewise [16: Art. 153 et seq.].

Finally, Art. 127.1 of law no 4727/2020 lays down that rights of use for radio frequencies cannot be restricted or withdrawn before expiration of the period for

[23] The corresponding procedure is decided upon by the Minister of Digital Governance [16: Art. 119.2].

[24] For more details on the procedures used for the granting of such rights, see Art. 119.2 of Law no 4727/2020 [16: Art. 119.2].

[25] By way of exception to these provisions, the Minister of Digital Governance may take account of the satellite market developments and particularities, and decide on the cession, renewal, amendment etc. of any rights of usage of orbital positions and associated frequencies, following an open tender procedure or by direct agreement.

[26] These conditions are laid down in more details in Annex I of the law [16: Annex I].

which they were granted, except where such thing is justified and in accordance with national law [16: Art. 127.1]. In this case, the restriction or withdrawal may be decided only where this is necessary to ensure the efficient use of radio frequencies or due to technical reasons, after paying compensation to the operator.[27] For the sake of completeness, it should be mentioned, as a final point, that Art. 231 clarifies that any provision contrary to law no 4727/2020, or to the relevant Part of law n° 4727/2020,[28] is repealed.

4.1.4 The Council for Space Policy, Established by Law n° 4508/2017

On the occasion of the debate on the national space law (in 2017), it was pointed out that in carrying out its purpose, the competent Ministry for space matters—i.e., Ministry for Digital Governance (former Ministry of Digital Policy, Telecommunications and Information [*Υπουργείο Ψηφιακής Πολιτικής, Τηλεπικοινωνιών και Ενημέρωσης*])—would also necessitate expert assistance from additional specialized entities (i.e., apart from the national space agency). Within this perspective, law n° 4508/2017 [20] created a particular body, namely a network of space specialists that would allow the ministry to benefit from a wider pool of expertise.

Following Art. 25.2 of law n° 4508/2017, which introduced a new paragraph to Art. 5 of law n° 4070/2012 [8] under the general title *Committees*, a new Council for Space Policy [*Συμβούλιο Διαστημικής Πολιτικής*] was created. More precisely, said council was to be set up after the adoption of a decision taken by the (former) Minister of Digital Policy, Telecommunications and Information [*Υπουργό Ψηφιακής Πολιτικής, Τηλεπικοινωνιών και Ενημέρωσης*].

The particular competences of the new authority were to provide information and advice services either on space-related matters, or on the elaboration of the national space policy, both to the competent minister and to other bodies entitled to decide on space matters.[29] Within this new consultative authority, the functions of president were to be exercised by the Secretary-General of Telecommunications and Posts,

---

[27] The terms of this compensation are established by joint decision of the Minister of Finance and the Minister of Digital Governance, following advice provided by the Hellenic Telecommunications & Post Commission [*Εθνική Επιτροπή Τηλεπικοινωνιών και Ταχυδρομείων*], [16: Art. 127.2]. A consultation procedure of the interested parties is also laid down, to allow the operator to defend its views [16: Art. 127.4].

[28] By using the expression "present Part of the law", reference is made to Part B entitled *Electronic communications Incorporation into Greek national law of the Directive (EE) 2018/1972 and other provisions*, comprising Art. 109 to 233, several of which were presented in this part of the study.

[29] Such as to the Hellenic Space Agency S.A., as it was stipulated in Art. 2.2 of the Decision of the Minister of Digital Policy, Telecommunications and Information n° 740, *on the Statute of the limited liability company named Hellenic Space Agency S.A. (ELDO S.A.)* [21].

while the members of the Council were determined in detail by Art. 25.2 of the law. *Inter alia*, the Director-General of Telecommunications and Posts and the Director of Satellite Services and Space Applications were to be members of the Council for Space Policy, in addition to representatives from relevant ministries, where appropriate [20: Art. 25.2].[30]

Several personalities regarded as contributing to the understanding of space matters were also entitled to participate in the Council for Space Policy, on an invitation from the Secretary-General of Telecommunications and Posts. These could be, in particular, representatives coming from the Hellenic Telecommunications & Post Commission [*Εθνική Επιτροπή Τηλεπικοινωνιών και Ταχυδρομείων/ EETT*], the academic community and/or the wider Greek space industry. Additionally, and where applicable, scientists and representatives from space-related bodies and institutions could be invited as well.

Moreover, it was stipulated that the Secretariat-General of Telecommunications and Posts would provide overarching support to the Council for Space Policy. By way of illustration, the Council for Space Policy could hold its meetings at the office of the Secretary-General of Telecommunications and Posts; the latter was at the same time entitled to determine matters to be discussed, or to invite the participation of other persons to the debates, taking into due account the level of security and confidentiality that would be required in each case [20: Art. 25.2].[31]

Finally, following the same line of reasoning, namely in order to effectively provide the Greek government with specialized knowledge and assistance, the legislator created the national space agency. As a result, said authority had specific powers to deal with space matters and to support the implementation of the adopted policy orientations.

4.2 The Greek National Space Agency

Section 4.1. clarified the role played by the competent ministry for space matters, together with its specialized divisions and services—as well as by the minister himself/herself—in regulating and coordinating space activity at the national level. However, the strategic and financial importance of space activity is constantly growing, and the number of States and non-State actors involved in space is steadily increasing. Against this background, issues are becoming more complex, and they tend to involve larger degrees of unpredictability; especially as a result of recent

[30] According to Art. 25.2 of law n° 4508/2017, representatives from relevant ministries are nominated by their ministers according to the subject matter discussed, together with their respective alternates.

[31] Research did not reveal further information on the more precise functioning of the *Council for Space Policy*.

economic and technological developments that took place in the context of space, space services and space applications.

Following on from that, the State apparatus had evidently to be enhanced with more specialized authorities, so as to better take into account the increasing importance and the complex aspects of many space operations, in addition to the growing number of private entities and stakeholders involved in this field. More precisely, inspired by the initiatives adopted in several other countries, Greece followed suit and created in 2017 the Greek national space agency.

4.2.1 Hellenic Space Agency S.A.—The Forerunner Created in 2017

In the late 2010s, Greece was still one of the few European countries that did not have a national space agency, despite being actively involved (for a long time already) in space activity. This point was underlined in the Explanatory Memorandum of law n° 4508/2017 [22: 1 point 2], to argue that the Greek State was therefore—possibly for this reason—unable to efficiently exploit its rights of use for radio frequencies or other space resources. At the same time, it was claimed that the country was (for the same reason) unable to optimize its participation in the ESA, and/or in any other international or inter-regional programs and activities.

Be that as it may, the investments made in the space sector had provided important long-term opportunities and allowed for economic growth in a new promising direction. Taking all this into account, a national space authority was thus created—under the name, *then*, of the Hellenic Space Agency—with the specific aim to achieve an adequate level of profitability on the public and private funds invested in said fields, coordinate efforts and provide assistance to the Greek State in this domain.

4.2.1.1 Setting Up the Hellenic Space Agency S.A.—Law n° 4508/2017

In response to the growing importance of space activities and the evolving demands of the Greek stakeholders, Law n° 4508/2017 [20] created the Hellenic Space Agency as the competent Greek authority for space matters. More specifically, provision was made in Chapter II [*Κεφάλαιο Β'*] of the law, titled *Creation of the Hellenic Space Agency* and comprising Articles 18 to 24, for the setting up of a specialized authority aimed at controlling and regulating space activity at the national level, as presented below.

On the basis of Art. 18.1, a limited liability company was created, under the name of Hellenic Space Agency S.A. [*Ελληνικός Διαστημικός Οργανισμός Α.Ε.*], and it was located in Athens[32] (hereafter, Hellenic Space Agency and/or the Agency). The Agency was operating under a specific regime: it was placed under the direct control and supervision of the Minister of Digital Policy, Telecommunications and Information [*Υπουργός Ψηφιακής Πολιτικής, Τηλεπικοινωνιών και Ενημέρωσης*]. Moreover, it was operating for public interest purposes under market-economy principles, subject to the provisions of law n° 3429/2005 [23] and—in a complementary way—to the provisions of Law n° 2190/1920 [24]. The term of the Company was set at thirty (30) years, starting from the publication of the Ministerial Decision which would lay down the company's statutes.[33]

Detailed rules on the principal objectives of the Hellenic Space Agency S.A. were provided for in Art. 18.4 of Law n° 4508/2017. In particular, the Agency was competent in key fields of space activity, such as in reference to policymaking and the management of space-related activities; the coordination and representation of the Greek space operators; and finally for activities aimed at research and knowledge development, including the ones aimed at technological innovation in the field of space, as outlined in the following paragraphs.

4.2.1.2 Competences of the Hellenic Space Agency S.A. Under Law n° 4508/2017

In reference to policymaking, the Agency was empowered to develop proposals for a space strategy—in addition to a rolling action-plan of space strategy—, in collaboration with the academic and research community, and the public and private sector, in order to set out the objectives, as well as any areas of cooperation and procedures needed to achieve these objectives [20: Art. 18.4.α]. As highlighted in Art. 24.3 of law n° 4508/2017, the proposal for a space strategy and the rolling action-plan had to be sent to the competent minister for approval (the law mentioned then the former Minister of Digital Policy, Telecommunications and Information [*Υπουργό Ψηφιακής Πολιτικής, Τηλεπικοινωνιών και Ενημέρωσης*]).

In the same vein, the Hellenic Space Agency S.A. was competent to provide expert and administrative assistance to the Greek State in reference to satellites and space matters, as well as to assist the competent ministry, by providing it with reliable and up to date data, assessments and advice on all these topics. Said duty included the provision of assessments and management advice on the rights and obligations of the State in outer space, the registration of space objects and the development of space systems [20: Art. 18.4.θ].

Furthermore, it was empowered to engage in national and international cooperation activities, concerning a wide range of space and space-related matters. In this context, it was entitled to establish cooperation, coordination, support

[32] The company's domicile could be changed by ministerial decision [20: Art. 18.1].

[33] The term of the Company could be amended or extended, as provided for in the Company's statutes [20: Art. 18.3].

and mobilization among space stakeholders, based on any form of synergy—such as between operators, agencies, and legal persons (governed by public or private law)—aimed at promoting, disseminating and making full use of the Greek space strategy. Within this purpose, the Agency was entitled to participate in related European and international organizations, or any other entities [20: Art. 18.4.β].

It was also competent to promote space projects and programs, participate therein as a coordinator for public bodies, and take part in the management of national projects and programs; such as mainly in those relating to scientific research, technology and telecommunications, and security [20: Art. 18.4.γ]. Moreover, it was qualified to develop and foster international involvement and collaboration, particularly through the participation of Greece in European and international organizations, initiatives, *fora* and activities on space topics; likewise, it was empowered to coordinate the involvement of national stakeholders; to promote the exchange of knowledge and know-how, i.e., with the aim to maximize the (contributory) participation of the country in any European and international space organizations, and finally to participate in the elaboration of programs aimed at meeting national—scientific, operational and economic—development needs, in which space and space resources play a key role [20: Art. 18.4.δ].

At the same time, the Hellenic Space Agency S.A. was entitled to contribute to the development and strengthening of any technology and space applications, services and ground-based infrastructure that can benefit to domestic industry and research. On this basis, it was responsible for providing support for the development of satellites, satellite systems, material, and equipment (as well as for remote-sensing activity), including by making full use of the country's memberships across European and international organizations [20: Art. 18.4.ε].

Importantly, the Hellenic Space Agency's mission was also to further improve the education and continuous training of students, young researchers and public officials, on space issues. In this vein, it was precisely empowered to participate in the planning and implementation of actions aimed at deepening the understanding of space activity and applications [20: Art. 18.4.στ], and to promote the standardization goals and the transfer of know-how and good practices to the benefit of the space sector; namely, to facilitate their utilization for space applications and services [20: Art. 18.4.ζ]. Ultimately, the Agency was competent to exploit the State's rights to use satellites and space objects—either for experimental or commercial purposes—, as well as to exploit any related services, assets, and resources, following an authorization granted by the competent minister [20: Art. 18.4.η].

4.2.1.3 Operating Rules of the Hellenic Space Agency S.A. Under Law n° 4508/2017

Law n° 4508/2017 established the operating rules for the functioning of the Hellenic Space Agency. Hence, according to Art. 18.5 of the law, the Agency's share capital (amounting to the equivalent of 60.000 Euros) consisted of 60 ordinary and registered shares, having each a nominal value of 1.000 euros; importantly, the whole amount was provided by the State. In exercising its share options, the only shareholder—namely, the Greek State—was represented in the Company's General Meeting by the competent minister (who was, then, the Minister of Digital Policy, Telecommunications and Information [*Υπουργός Ψηφιακής Πολιτικής, Τηλεπικοινωνιών και Ενημέρωσης*]) and by the Minister of Finance [*Υπουργός Οικονομικών*], or by any other persons authorized by them.

Following on from that, the particular features of the Agency's statutes had to be laid down in a decision enacted by the competent minister—the law made reference, then, to the Minister of Digital Policy, Telecommunications and Information [*Υπουργός Ψηφιακής Πολιτικής, Τηλεπικοινωνιών και Ενημέρωσης*]—, which was indeed duly adopted and subsequently published in the Greek Government Gazette [21], as it was stipulated in Art. 18.6 of law n° 4508/2017. As a result, the statutes regulated (closely following Art. 18.6 of the Greek space law) issues relating to matters of limited liability companies, and the Agency was registered in the General Commercial Registry (GEMI) [*Γενικό Εμπορικό Μητρώο (Γ.Ε.ΜΗ.)*]. As a general rule, law n° 4508/2017 established that the Agency's statutes and any relevant provision could be amended and codified following a decision of the General Assembly. Importantly, in fulfilling its objectives, and subject to approval by the competent minister, it was explicitly stipulated that the Agency could create subsidiaries and branches. In the same vein, it was established that it could collaborate with Greek and foreign companies involved in space activities; such as by creating consortia, participating as an associate or a shareholder in any other entity involved in this field and by establishing joint ventures.

Regarding its means of operation, law n° 4508/2017 established (and the rule was also reiterated in Art. 7.2 of the statutes) that the Hellenic Space Agency S.A. resources would come, *first*, from the exploitation of any State's rights to use space objects, satellites, orbital positions, associated frequencies etc. which are granted for use to the Agency by the competent minister [20: 18.7.α]. In addition to that, further resources could consist of sponsorships, funding and other resources coming from State-owned undertakings and companies, EU funds and other international organizations [20: 18.7.β]; the provision of services and/or the transfer of rights to use any scientific and research works [20: 18.7.γ], resources coming from donations, successions, revenues derived from the exploitation of its share capital etc. [20: 18.7.δ], as well as from funding arising from private and corporate sponsorships [20: 18.7.ε]. At the same time, law n° 4508/2017 laid down specific rules governing the management and conditions of exploitation of the Agency's assets [20: Art. 18.8 and Art. 18.9] all of them reiterated, again, in Art. 7.2 of the statutes.

In this context, it was expressly stipulated that the Hellenic Space Agency S.A. could receive public funds from the Program of Public Investment [*Πρόγραμμα Δημοσίων Επενδύσεων (ΠΔΕ)*] of the State Budget. Still, in such a case, a joint decision was required, signed by the competent minister for space matters (then, mention was made to the Minister of Digital Policy, Telecommunications and Information [*Υπουργό Ψηφιακής Πολιτικής, Τηλεπικοινωνιών και Ενημέρωσης*]), the Minister for Finance, Economy and Development [*Υπουργός Οικονομικών, Οικονομίας και Ανάπτυξης*] and any other competent minister. At the same time, it was established that EU State Aid Law would apply in this case [20: Art. 18.10 and Art. 22.2], although military applications and national safety issues were excluded from the scope of this provision [20: Art. 18.10]. Importantly, regarding space programs and applications serving national defense purposes, it was also mentioned that the Ministry of National Defense would be granted administrative and operational autonomy from the Agency, while the Minister of National Defense was given the right to attend the Board meetings upon invitation—without, though, any voting rights—, in case that national defense and national security matters were being discussed [20: Art. 18.11].

4.2.1.4 Management Bodies of the Hellenic Space Agency S.A.: Law n° 4508/2017

Law no 4508/2017 stipulated that the governing bodies of the Hellenic Space Agency S.A. were the *General Assembly* and the *Administrative Board*, comprising the Chairman, the Executive Director, and the Board Members [20: Art. 19]. All relevant rules were developed in Art. 3 et seq. of the statutes, and are briefly presented below.

More precisely, the *General Assembly* was established as the supreme body of the Agency [20: Art. 19.2]. As the principal organ of the Hellenic Space Agency S.A., it was empowered to decide on any matter explicitly referred to in the statutes[34] and in law n° 2190/1920 [24]; any other issues had to be addressed in accordance with a particular procedure that was expressly provided for in law n° 4508/2017. In this context, the General Assembly was precisely competent to appoint the Administrative Board, for a period of five years [20: Art. 19.5]; although the initial Administrative Board, which was to oversee operations until the creation and staffing of the Agency were finalized, was appointed by the competent minister—the Minister of Digital Policy, Telecommunications and Information ["*Υπουργό Ψηφιακής Πολιτικής, Τηλεπικοινωνιών και Ενημέρωσης*"], then—, who at the same time decided on the manner of its functioning.

The authorities responsible for the Agency administration, and for the implementation of the General Assembly's decisions, were the *Administrative Board* and its *Executive Director*. All matters relating to the extent of their powers and to the overall decision-making process were regulated in Art. 4 to 6 of the Agency statutes, and in line with Law n° 4508/2017.[35]

More precisely, the *Administrative Board* composition was laid down in Art. 19.4; practically, it consisted of seven (7) members, comprising—in addition to the Chairman and the Executive Director—two Board Members appointed by the

[34] Art. 3 of the statutes simply reiterated the rules and competences established in law no 4508/2017.

[35] Further details on the establishment, the functioning, and the powers of the governing and executive bodies of the Hellenic Space Agency S.A. were laid down in Law n° 4508/2017, in Art. 19.3 to 19.7. The statutes mainly reiterated the substance of these rules.

(former) Minister of Digital Policy, Telecommunications and Information [*Υπουργό Ψηφιακής Πολιτικής, Τηλεπικοινωνιών και Ενημέρωσης*]; one Board Member appointed by the Minister for Finance; one Board Member appointed by the Minister of Education, Research and Religious Affairs and, finally, one Board Member designated as the representative of all employees. The law additionally determined which rules would apply in case the above authorities failed to propose candidates for appointment as Administrative Board Members, along with their respective alternates, within ten (10) days of receiving the relevant request [20: Art. 19.4].

The specific powers of the *Administrative Board* were clearly defined in Art. 19.5 of law n° 4508/2017. In particular, it was established that the Board would be entitled to decide on all matters relating to the management of the Agency, apart from issues that were within the competence of the General Meeting. Interestingly, where necessary, the Administrative Board was also given the possibility to delegate the handling of particular issues to external experts, renowned for their specialized knowledge, skills and experience on space matters; in such an event, their remuneration would be determined by the same decision of the Administrative Board appointing them. Finally, the powers and qualifications of key members of the Administrative Board, such as mainly those of the Chairman and the Executive Director, were further clarified in the law [20: Art. 19.3, 19.6 and 19.7] in conjunction with Art. 5 and 6 of the statutes [21].

By way of illustration, according to Art. 19.3 of the Greek space law, the *Executive Director* was competent to supervise the Agency Directorates and take all the necessary implementing decisions, as also detailed further in the statutes.

4.2.1.5 Internal Functioning of the Hellenic Space Agency S.A.: Law n° 4508/2017

To complement the statutes, clarifications regarding the internal functioning of the Agency—e.g., on internal governance, security and staff classification, personal data and classified information protection, or control matters—, had to be agreed upon by the Administrative Board following adoption of the Agency's Internal Rules [*Εσωτερικοί κανονισμοί*], as established in Art. 20.1 of law n° 4508/2017.[36] On this basis, the Administrative Board was entitled to decide on conditions and procedures for the award and implementation of agreements aimed at the supply of goods and/or the provision of services; the conduct of any relevant studies, and/or the award and implementation of public work contracts. In addition to that, it was empowered to adopt measures on the administration and management of the Agency

[36] According to Law no 4508/2017, the Company's Internal Rules could cover questions such as the organization, the functioning, the structure, and the powers of the Hellenic Space Agency S.A. Divisions; the number of positions for each service and each category of personnel; the number of experts, legal advisers etc. needed; the recruitment and promotion criteria and procedures etc. [20: Art. 20.2]

assets, such as goods purchases, exchanges and sales of movable and immovable property, or lease contracts.

Given the importance of the Agency as part of the State apparatus, law no 4508/2017 also laid down the rules applying to its personnel [20: Art. 21] and determined the privileges and tax exemptions accorded to the authority. Furthermore, it explicitly stipulated that the Administrative Board would provide any information related to its functioning, at the official request of the (former) Minister of Digital Policy, Telecommunications and Information [*Υπουργό Ψηφιακής Πολιτικής, Τηλεπικοινωνιών και Ενημέρωσης*], [20: Art. 24.4]. Finally, it mentioned that the Administrative Board could be invited to appear before committees of the Parliament, to provide information on any space issue, without in any case prejudice to matters of national security that are related to military applications and space resources management [20: Art. 24.5]. At the same time, the Agency was explicitly placed under the sustained supervision of the competent minister, who was then the Minister of Digital Policy, Telecommunications and Information [*Υπουργός Ψηφιακής Πολιτικής, Τηλεπικοινωνιών και Ενημέρωσης*], [20: Art. 24.2, Art. 24.3 and Art. 24.4].

In case of dissolution of the Agency,[37] Art. 23 of the Greek space law laid down that the liquidation proceedings would be conducted in accordance with the provisions contained in law n° 2190/1920 on Public Limited Liability Companies [24]. Nevertheless, specific issues—such as those concerning the Agency's personnel or (tangible and non-tangible) assets—, could still be decided by joint decision of the (former competent) Minister of Digital Policy, Telecommunications and Information [*Υπουργός Ψηφιακής Πολιτικής, Τηλεπικοινωνιών και Ενημέρωσης*], together with any other minister deemed competent for each specific matter [20: Art. 23].

A couple of years later, following adoption of the Presidential Decree n° 81/2019 [13] released on 8 July 2019, the Hellenic Space Agency S.A.—which had until then been under the authority of the (former) Minister of Digital Policy, Telecommunications and Information [*Υπουργός Ψηφιακής Πολιτικής, Τηλεπικοινωνιών και Ενημέρωσης*]—, was eventually placed under the jurisdiction of the newly created Ministry of Digital Governance [*Υπουργείο Ψηφιακής Διακυβέρνησης*], [13: Art. 1 para. 2.1].

After a month, the (then existing) Hellenic Space Agency S.A.—created by Art. 18 of law n° 4508/2017—was repealed without undergoing liquidation,[38] pursuant to Art. 60 of law n° 4623/2019 adopted on 9 August 2019 [14]. Simultaneously, a new national space authority assumed the functions of the Hellenic Space Agency, as its universal legal successor: the Hellenic Space Center [*Ελληνικό*

[37] Control, supervision, and transparency rules were established in Law no 4508/2017, Art. 24, laying down that in this field, Law no 3429/2005 [23] would apply.

[38] The same provision mentioned that the Hellenic Space Agency S.A. is deleted from the General Commercial Registry (GEMI) [*Γενικό Εμπορικό Μητρώο (Γ.Ε.ΜΗ.)*], [14: Art. 60.2].

Κέντρο Διαστήματος], which currently serves as Greece's competent authority for space matters.

4.2.2 *The Hellenic Space Center*

The space authority that is at present competent to act, at the national level, for the implementation of space law and policy matters, is the *Hellenic Space Center* [*Ελληνικό Κέντρο Διαστήματος*] created under Art. 60 of law n° 4623/2019 [14] (hereafter, Hellenic Space Center, or the Center). The Hellenic Space Center was established as a non-profit-making legal person governed by private law [14: Art. 60.1], serving as the universal legal successor to the former *Hellenic Space Agency S.A.* and entitled to exercise all the powers and competences of its forerunner [14: Art. 60.3].

Against this background, any type of contract launched and/or concluded by the former Hellenic Space Agency S.A.—namely, launched and/or concluded before the entry into force of law n° 4623/2019, and/or still under execution after the new law was adopted—was being continued in the name of the new authority. As a result, neither the new Center, nor any other contracting party or third person, was entitled to resolution or resiliation of said agreements for this particular reason. In addition to that, all the existing activities and programs, or powers and financing granted to the (former) Hellenic Space Agency S.A. were automatically transferred to the new authority, while trials of pending cases were being continued under the name of the Hellenic Space Center [14: Art. 60.3].

4.2.2.1 Functioning of the Hellenic Space Center under Law n° 4623/2019

The *Hellenic Space Center* is situated in Athens, Greece. According to Art. 60.4 of law n° 4623/2019, the location of its offices may be changed following a decision by the Minister of Digital Governance [*Υπουργός Ψηφιακής Διακυβέρνησης*], [14: Art. 60.4]. For the sake of precision, Art. 60.5 underlines that the Center forms part of the broadly defined public sector. Notably, it carries out a function to the benefit of the public interest, and it is governed by the relevant provisions of law n° 4623/2019 as well as by its Internal Rules; furthermore, it enjoys exemption from taxes and such other privileges applicable to public entities, as determined in the same provision coupled with Art. 60.6.[39]

[39] Law n° 4623/2019, Article 60.5 reiterates the rule that tax exemptions measures must be applied in accordance with EU State aid law, with the exception of matters related to military applications and national safety issues (the same rule was established in Law n° 4508/2017, Art. 18.10 [20]).

The Hellenic Space Center is placed *expressis verbis* under the direct responsibility of the Minister of Digital Governance [*Υπουργό Ψηφιακής Διακυβέρνησης*], [14: Art. 60.7] which is testament to the growing importance that the Greek State attaches to space matters and activity. In this context, the new authority's competences are enumerated in Art. 60.8 which reproduces, *verbatim*, the provision of Art. 18 of law no 4508/2017 [20][40] that had established the powers of the former Hellenic Space Agency S.A. (*nb.* as the same rules on the entity's competences apply, the presentation of the provisions concerned—initially laid down in Art. 18 of law no 4508/2017—will not be duplicated here).

Importantly, however, Art. 60.9 of law n° 4623/2019 expanded the financial resources available to the new authority. Hence, these resources now encompass any type of funding or subsidy from the public budget [14: Art. 60.9.α]; any type of financing, contribution or grant coming from institutions or bodies of the public and/or private sector, as well as from the EU and other European or international organizations [14: Art. 60.9.β]; revenues generated from the exploitation of the State's rights to use space objects, satellites, orbital positions and associated frequencies allocated to the Hellenic Space Center by the Minister of Digital Governance [14: Art. 60.9.γ] etc.[41]; income derived from the provision of specialized services and the execution of projects on behalf of third parties [14: Art. 60.9.στ], and proceeds from the management of tangible and non-tangible assets or loans [14: Art. 60.9.ζ]. Notably, it was additionally established that in the event of the dissolution of the Hellenic Space Center, its assets—including all of its rights and obligations—shall automatically revert to the Greek State.

At the same time, Art. 60.10 of law n° 4623/2019 reiterated the rule laid down in law n° 4508/2017, Art. 18.11, in reference to space programs and applications serving national defense purposes.[42] Hence, the Ministry of National Defense retains control, through administrative and operational autonomy, over any such space programs and applications. As a result, it is granted the power to participate—without voting rights and only when convened—in the meetings of the Center's Governing Board, whenever matters related to national defense and security are debated.

---

[40] The only difference is the new competent Minister and Ministry being mentioned, where appropriate, in Law n° 4623/2019.

[41] The resources mentioned in Art. 60.9 of law no 4623/2019 include, in addition to those precisely referred to in said article (e.g., Art. 60.9.α and 60.9.β), also the ones enumerated in Art. 18.7 of law no 4508/2017.

[42] In this case, the Ministry of National Defense is granted administrative and operational autonomy and may attend the meetings of the Hellenic Space Center's Management Board, upon invitation (without any voting rights), in case that national defense and national security matters are debated.

4.2.2.2 Management Bodies of the Hellenic Space Center

Finally, law n° 4623/2019 established detailed rules concerning the Hellenic Space Center financial management and control,[43] as well as concerning its administration. More precisely as regards the governance of the new authority, Art. 60.12. stipulated that its principal decision-making body would be the Governing Board [*Διοικητικό Συμβούλιο*], appointed by the Minister of Digital Governance [*Υπουργό Ψηφιακής Διακυβέρνησης*][44] and comprising seven (7) members. According to Art. 60.12, its members are the Chairman, the Chief Executive and five (5) Board members.

In this context, the Chairman's powers were specifically laid down in Art. 60.13[45] of law n° 4623/2019: he/she is the authority competent to supervise the Hellenic Space Center's work; monitor its functioning and report on the Center's activity to the Board. As clearly underlined in the law [14: Art. 60.13], he/she was established as its higher executive organ.

On a parallel level, the Governing Board is responsible for the management of the Hellenic Space Center, as well as for achieving its objectives, on the basis of Art. 60.15 of law n° 4623/2019. Hence, it was empowered to decide on questions related to the administration of the Center's assets; the management and representation of the new authority, and take all the necessary measures so as to ensure the full realization of its objectives. Overall, the Board decides on matters concerning the administration of the new authority, except for those that fall within the exclusive competence of the Minister of Digital Governance [*Υπουργό Ψηφιακής Διακυβέρνησης*]. In doing so, the Board decides by a majority of the votes cast (in case of equality, the Chairman shall cast the deciding vote).

Moreover, to ensure the soundness of the management decisions taken, Art. 60.14 enumerated the required professional competencies and qualifications to be eligible as a Member of the Board, whereas additional eligibility criteria (such as, conflicts of interest and/or incompatibilities) may be determined in the Internal Rules. Needless

[43] Law no 4623/2019, Art. 60.11 stipulates that the Hellenic Space Center must keep records and supporting documentation for the purpose of controls; this information must be available to the competent authorities, upon their request. The monitoring and inspections of the financed projects are conducted in order to supervise the ongoing activities and to ensure that the funds are invested in accordance with the objectives of law n° 4623/2019 and the new authority decisions. In addition to that, the Hellenic Space Center must produce an Annual Report, containing a financial report (that must be submitted to the Minister of Digital Governance) as well as a report on its past activities, to be published on its official website. The same article provides further details on the content of these reports, and on the auditing of its financial management [14: Art. 60.11].

[44] The current members of the Governing Board were appointed by decision of the Minister of Digital Governance adopted on 18 January 2020 [25], for a period of five years starting from the publication of the decision.

[45] According to law n° 4623/2019, Art. 60.13, the Chairman of the Board is the competent authority to convene ordinary (at least, 12 times a year) and extraordinary meetings of the Board, to draw up its agenda and direct its work. The same article provides further detailed rules governing the convening and the functioning of the Board, such as on the appointment of members to fill the vacancies that may occur, rules on its functioning during this period etc.

to say that in implementing its objectives or wherever it is necessary, the Board—and the Hellenic Space Center, likewise—may also engage external well-known space experts.[46]

Eventually, the Hellenic Space Center had to adopt its own Internal Rules (i.e., rules of procedures) within one year of the date of publication of law n° 4623/2019, and to submit those for approval to the Minister of Digital Governance [*Υπουργό Ψηφιακής Διακυβέρνησης*]. In this regard, Art. 60.20 stipulated that the Internal Rules must address specific issues, such as those related to the functioning and working procedures of the Center, reiterating the content of law no 4508/2017 [20: Art. 20.1 and 20.2].[47]

As a final note, it is worth mentioning that the Hellenic Space Center was recently included—on the basis of Art. 23 of law n° 5039/2023 [26]—within the category of Technology Institutions and Bodies enumerated in law n° 4310/2014 [6] which regulates, among other things, matters of research, technological development and innovation. Following on from that, its functioning is now also governed by law n° 4310/2014 for the topics it addresses.

4.3 Other Institutions and Bodies Impacting Space Matters

At present, the National Space Policy is co-developed and implemented by the Ministry of Digital Governance—as assisted by its specialized services—and the Hellenic Space Center, within the framework of the European and international context. Put differently, in giving effect to its obligations under international space law, the Greek State may rely—in addition to the competent minister—on several State bodies and authorities being specialized on space matters and comprising numerous technical, administrative and legal experts. In principle, they form part of said ministry or may have some degree of autonomy, for a greater flexibility (though remaining under its responsibility) as explained above under Sects. 4.1 and 4.2.

Notwithstanding, in practice, the administration also looks forward to establishing a complementary though systematic cooperation with scientific

[46] Pursuant to law n° 4623/2019, Art. 60.15, the Board may collaborate with experts outside the Hellenic Space Center staff who possess extensive experience and expertise relevant to the tasks assigned to them; their remuneration is decided by decision of the Board. In addition to that, Art. 60.16 mentions that the Hellenic Space Center enjoys a high degree of flexibility to hire staff through various contracting arrangements, including private-law employment contracts (e.g., fixed-term or open-ended contracts); a remunerated mandate agreement; a works contract, etc. As a derogation, the new authority may also hire up to ten (10) well-known experts to assist it in achieving its goals, precisely as regards the monitoring of space activities and space objects, etc. [14: Art. 60.17 to 60.19].

[47] It was further mentioned that until the adoption of the new authority Internal Rules, decisions in such matters would be made by the Governing Board of the Hellenic Space Center [14: Art. 60.20].

institutions specialized in space topics, due to the characteristic complexity of space and space-related matters. The most influential ones are mainly non-profit entities that are active in space-related fields (Sect. 4.3.1). Importantly, though, the Greek government takes additionally into consideration the Greek industry concerns and interests when developing space policy and/or legislation, especially in case that policymaking in the field of space is likely to affect the business climate (Sect. 4.3.2).

4.3.1 Assistance Received from Space-Related, Non-profit Institutions

As regards initially the category of non-profit entities, a distinction should be made between *first*, the ones attached to the ministry—whose purpose is to *directly* provide the Greek State with expert advice on space-related matters—and second, the academic and research institutions. Quite characteristically, the second category seems to serve as an *indirect* source of expert advice to the competent State authorities (namely, although that was not their primary purpose) by disseminating the most up-to-date knowledge about space matters as well as related technical information.

4.3.1.1 The Greek National Committee for Astronomy—Leg. Decree n° 3725/3-8-1957

A typical example of the first category is the Greek National Committee for Astronomy [*Εθνική Αστρονομική Επιτροπή*], which was established in 1957 as the primary body empowered to provide independent expert opinion to the administration, on matters related to astronomical and astrophysical research [27]. Its creation followed the adoption of legislative decree n° 3725/3-8-1957 [28].

Back then, the National Committee for Astronomy was created with the aim to promote and coordinate the work carried out in astronomy, mainly in reference to matters requiring international cooperation, as well as to liaise with the International Astronomical Union [28: Art. 1 para. 1]. It was placed under the supervision of the Ministry of National Education and Religious Affairs [*Υπουργείο Εθνικής Παιδείας και Θρησκευμάτων*], and its composition was determined in detail by Art. 1 para. 2 of its founding legislative decree.

However, following Art. 1 para. (β) of the royal decree adopted on 20 June 1958, which was regulating the organization and operation of the National Committee for Astronomy [29], the provision of scientific opinion and advice on astronomical matters, to the Minister for Education [*Υπουργόν της Παιδείας*], was explicitly added to its competences. It remains, to this day, the official independent advisory body to the Greek government, entitled to inform it on the views of the scientific community in the fields of astronomy and astrophysics [30].

4.3.1.2 The Hellenic Astronomical Society

From a different but complementary perspective, the Hellenic Astronomical Society (Hel.A.S.) [*Ελληνική Αστρονομική Εταιρεία*] was established in 1992, with the aim to provide a platform for the discussion of space-related matters of importance, and to also promote collaboration among experts, researchers and scientists in this field. As an academic institution, it is principally focusing on the collection of data on space and space-related issues, with the purpose of examining present and future challenges.

The first discussions on the establishment of a national astronomical society date back to 1982, and took place in the context of the XVIII General Assembly of the International Astronomical Union, in the Greek city of Patras. Almost a decade later, the constitution of the Hellenic Astronomical Society was finally presented to—and approved by—the participants of the first Hellenic Astronomical Conference, held in Athens (in September 1992). Said constitution, amended in 2002 and 2021, was legally approved by decision n° 90/2022 of the Court of Athens [31] (up to now, Hel. A.S. has 288 members, two thirds of which are associated with Greek academic institutions [32]).

According to Art. 2 of its Constitution, the aim of the Hellenic Astronomical Society (Hel.A.S.) is to encourage the development and promotion of astronomy *lato sensu* in Greece, and to provide support and advice to its members. Within this purpose, Hel.A.S is empowered to facilitate discussions on astronomy and to improve the dissemination of astronomy-related information to the general public and to any person concerned [31: Art. 2 para 2]; which may very well include the Greek State, as well as any State authorities and bodies competent to decide on space matters.

The Hellenic Astronomical Society is officially recognized as affiliated member of the European Astronomical Society, and it has established links with other international astronomical societies. Against this background, one may argue that its contribution to national policymaking is, in a sense, an indirect one, since it is principally collaborating with the Greek National Committee for Astronomy, which is the official advisory body to the Government for space-related issues [33].

4.3.2 Coordinating Bodies of Profit-Making Entities Having an Influence

In addition to the assistance provided by the above entities, stronger linkages are being developed between the private sector and State authorities, through efforts aimed at fostering enhanced cooperation with the coordinating bodies (and consortia) of the former. In truth, the increased participation of private stakeholders in space—within the context of innovative public-private financing mechanisms— is not without posing its own set of challenges (i.e., private actors aim to operate in space under market conditions, minimizing the integration of public interest or policy objectives into their managerial decisions; however, their flexibility and ability to leverage advanced technologies sometimes enable them to pioneer innovative approaches).

Practically speaking, they are becoming more and more involved in decision-making, through direct or indirect interaction with State authorities and society, even while remaining focused on their own priorities. In this context, Greek space companies have created associations or partnerships to coordinate efforts at the national level, advance their financial interests, and notably, represent their views on space matters before State bodies.

4.3.2.1 The Voice of the Greek Space Industry Association

The Hellenic Association of Space Industry (HASI), hereafter HASI or Association, [*Ένωση Ελληνικών Βιομηχανιών Διαστημικής Τεχνολογίας & Εφαρμογών* / (ΕΒΙΔΙΤΕ)] was founded on 22 September 2008, as a non-profit legal person under private law [34],[48] by a decision of the General Meeting based on Art. 1 and 6 of its Statute. The HASI was established for a period of twenty-five (25) years which can be extended indefinitely, and its objective is to promote the competitiveness of the Greek industry in the fields of space technology and space applications. In this context, it plays a key role in enhancing the exchange and dissemination of relevant information to promote and efficiently support all its members [35: Art. 2 and Art. 6].

More specifically, Art. 2 of the Statute lays down that HASI activities comprise the promotion and assistance of the Greek space (technology and applications) industry both at the national and international level. On this basis, it aims at ensuring effective communication, interaction and coordination between its members[49] and at

[48] The highest decision-making body within the Hellenic Association of Space Industry (HASI), being responsible for approving the major decisions affecting its functioning, is the General Assembly. Its composition and mode of operation are laid down in detail in Art. 13 of the HASI Statute [35]. In addition to that, the HASI Administrative Board decides on more practical matters, as established in Art. 8. Said body is elected by the General Assembly (under conditions of quorum and majority referred to in Art. 8 para.5) to serve a mandate of two (2) years; its members may be re-elected without limit (Art. 8 para. 3 and 4). Its competences, as well as its composition and functioning are described in detail in Art. 8 and 9 of the Statute [35].

[49] Pursuant to Art. 3 of the HASI Statute, members comprise natural or legal persons, who must respect the criteria that are set out in Art. 4 and 5 of said document; they are divided into full members, associate members and probationary members. In particular, its members may be legal persons carrying on space-related activities [35: Art. 4] in areas such as "electronics; advanced composite and metallic materials and structures; electronic satellites; specialized software support

promoting the further development of Greek space technology and high-tech sectors [35: Art. 2 para. 1]. Moreover, HASI aims at developing networks for communication and cooperation between its members, to enhance collaboration with relevant bodies or associations of third countries [35: Art. 2 para. 2]. Interestingly, its objective is also to promote cooperation between the national space industry and any domestic research center, institution and/or university involved in space-related fields [35: Art. 2 para. 4]; and, finally, to promote the optimization of the Greek participation in the ESA or in any other similar organizations, in order to seek new opportunities for industrial cooperation and synergies potentially resulting from any of the aforementioned participations [35: Art. 2 para. 3].

In reference to policymaking, given its objectives to facilitate the dialogue between its members and to also promote the Greek space industry interests,[50] HASI may set up any advisory or expert committee to address matters of common interest relating to space activity. In the same vein, it is additionally competent to provide guidance and advice to the Greek Parliament, in order to support policymaking in the field of space [35: Art. 2 para. 5]. In particular, it is empowered to carry out reports, studies and analyses on any space (and space related) issues of key interest, with the purpose of publishing them or submitting them to any competent administrative and/or legislative authority [35: Art. 2 para. 6]. Given the importance of the HASI members[51]—and the work carried out by the

satellite systems; MEMs technology sensors; ion motor development; collection, processing and development of remote sensing data applications; mechanical ground support equipment, etc.". To acquire full membership, a legal person is required to perform all or part of the functions related to its (space-related) business activity in Greece. The legal person's equipment, infrastructure and personnel, being used to carry out the company's activity, must be located in Greece, likewise. Associate members may comprise natural persons or legal entities acting in related fields, such as academic and research institutions, law firms and consultancy firms, as well as startups. They may—at their request—become full members, following a decision from the Administrative Board [35: Art. 5.1].

[50] Namely, at a practical level, the HASI aims at contributing to the best use of financing programmes related to space research and development, to the benefit of the Greek industry that is active in space technology and space applications. In correlation to that, it is also empowered to promote and disseminate the results of space-related research activities [35: Art. 2 para. 7 and 9]. Furthermore, it promotes activities and initiatives in the fields of space, such as telecommunication, remote-sensing, navigation and multimedia applications [35: Art. 2 para. 8] and it is competent to provide assistance regarding the education and training of its members' personnel, or on company development issues [35: Art. 2 para. 10]. Finally, the HASI is empowered to follow-up advances in the fields of space technology and space applications [35: Art. 2 para. 11].

[51] As highlighted by the entity, "HASI members are high-tech companies, employing over 2500 employees, of which the vast majority constitutes highly educated personnel in space technology and its applications. HASI members currently accumulate Greece's research and manufacturing person power in the space technology and applications sector. From 2003 up to date, HASI members have successfully participated in over 50 space technology and applications programs, involving ESA as well as other international programs. HASI members had a cumulative budget of more than 185 M€" [34].

Association[52]—, its increasing participation in the policymaking context is both valuable and essential to ensure a comprehensive exchange of views on space-related topics at national level.

4.3.2.2 Practical Participation of the Greek Space Companies

Finally, given the strategic importance of the industry for policymaking in the field of space, and the resulting attention given to the involvement of private stakeholders in the definition, co-drafting and implementation of space policy goals, the participation of such entities is being steadily encouraged. In line with this reasoning, a growing role is played *inter alia* by the Space Technologies and Applications Cluster, known as si-Cluster [36].

More precisely, the si-Cluster was created in 2009, as a joint initiative of Corallia, a Unit of the Research Centre Athena, and the Hellenic Association of Space Industries (HASI). It is described as "an industry-led and user-driven innovation cluster active in the field of Space Technologies and Applications in Greece. It brings together more than 70 Greek businesses, academic and research centers and other private and public sector organisations and the civil society" [37]. As part of its mission, which is "to place Greece on the global map of this highly developing and competitive market, based on the country's rich scientific potential and build capacities to attract high-impact research, innovation and commercial activities that will benefit the society" [37], the si-Cluster also aims to assist in Greece's broader efforts in this field, by defining precise strategic goals. These goals include establishing a coordinated National Strategy in the field of the space industry, a National Space Program, National Space Infrastructure and Facilities, fostering bilateral relations and cooperation with other European countries, and increasing Greek participation in space programs [36].

In addition to the above, "Every year, the si-Cluster conducts an inventory of the Greek space ecosystem, value chains, activities and technologies in Greece, leading to the Greek Space Catalogue and a bottom-up contribution for the Greek Space Strategy document and implementation plan" [36].

Overall, the si-Cluster aims at bringing together private and public stakeholders in the field of space technology and applications, with the purpose to elaborate an efficient framework, capable of developing and attracting high impact research, innovation and business activities. In this context, it supports the development of a competitive Greek space industry, and encourages the full use of services provided by space technology, while also providing key assistance for the development and implementation of space-related programs for different ministries and governmental

[52]Up to now, the HASI has organized a wide range of activities. Additionally, its members have successfully participated in numerous space missions, such as SOLAR ORBITER, EUCLID, JUICE, PLATO, EXOMARS ROVER MISSION, SENTINEL MISSIS I AL, and ESA Technology Programs [34].

entities (such as the Ministry of Defense, or the Ministry of Environment, Energy and Climate Change [38]).

References

1. Hellenic Parliament. *The constitution of Greece as revised by the parliamentary resolution of November 25, 2019 of the IXth Revisionary Parliament*. Accessed from https://www.hellenicparliament.gr/en/Vouli-ton-Ellinon/To-Politevma/Syntagma/
2. Mavrias, K. (2021). *Constitutional law* (6th ed.). P.N. Sakkoulas S.A.
3. European Union. *Law n° 4622/2019 entitled Executive State [Επιτελικό Κράτος]: Organization, functioning and transparency of the state, the governmental bodies and the central public administration*. Published at the Greek Government Gazette A' 133/07.08.2019, pp. 3223 et seq.
4. European Commission. *Ministry of digital governance [Υπουργείο Ψηφιακής Διακυβέρνησης]*. Accessed from https://mindigital.gr/
5. *Law n° 2953/19.10.2001 on the ratification of the agreement between the government of the Hellenic Republic and the European Space Agency concerning space cooperation for peaceful purposes*. Published at the Greek Government Gazette A' 249/22.10.2001, p. 3479. et seq.
6. *Law no 4310/2014 titled Research, technological development and innovation and other provisions*. Published at the Greek Government Gazette A' 258/08.12.2014, p. 7867 et seq.
7. *Presidential Decree 65/27.06.2011*. Published at the Greek Government Gazette A' 147, p. 3125 et seq.
8. *Law no 4070/2012 titled Regulatory framework for electronic communications, transports, public works and other provisions*. Published at the Greek Government Gazette A' 82/10.04.2012, p. 2131 et seq.
9. *Presidential Decree 24/27.01.*2015. Published at the Greek Government Gazette A' 20, pp. 173 et seq.
10. *Presidential Decree 123/4.11.2016*. Published at the Greek Government Gazette A' 208, p. 9333 et seq.
11. *Law no 4463/2017 on the measures for the reduction of costs for the deployment of high-speed electronic communications—Harmonization of national legislation with directive 2014/61/EU and other provisions*. Published at the Greek Government Gazette A' 42/30.03.2017, p. 479 et seq.
12. *Presidential Decree 82/10.08.2017*. Published at the Greek Government Gazette A' 117, p. 2103 et seq.
13. *Presidential Decree 81/8.07.2019*. Published at the Greek Government Gazette A' 119, p. 3181 et seq.
14. *Law no 4623/2019 titled Arrangements related to matters inherent to the Ministry of Interior, eGovernment provisions, pension arrangements and other urgent issues*. Published at the Greek Government Gazette A' 134/09.08.2019, p. 3295 et seq.
15. *Presidential Decree 40/15.04.2020*. Published at the Greek Government Gazette A' 85, p. 1393 et seq.
16. *Law no 4727/2020 on the Digital Governance (Incorporation into Greek Law of Directive (EE) 2016/2102 and Directive (EE) 2019/1024) – Electronic Communications (Incorporation into Greek Law of the Directive (EE) 2018/1972) and other provisions*. Published at the Greek Government Gazette A' 184/23.09.2020, p. 9623 et seq.
17. Kathimerini. (2019, April 09). *Greece signs cooperation agreement with NASA at Space Symposium*. Accessed July 2024, from https://www.ekathimerini.com/news/239377/greece-signs-cooperation-agreement-with-nasa-at-space-symposium/

18. US Ambassy & Consulate in Greece. (2019, April 19). *Ambassador Pyatt's remarks at the "Hellas to the Moon" Forum.* Accessed July 2024, from https://gr.usembassy.gov/ambassador-remarks-04192019/
19. *Constitution and Convention of the International Telecommunication Union*, 22 December 1992, UNTS 1825, 1826 (entered into force on 1 July 1994).
20. *Law no 4508/2017 on the authorization of space activities - Registration in the National Registry of space objects - Establishment of the Hellenic space agency and other provisions.* Published at the Greek Government Gazette A' 200/22.12.2017, p. 4411. et seq.
21. *Decision of the Minister of Digital Policy, Telecommunications and Information n° 740, dated 6 March 2018, entitled statue of the limited liability company named Hellenic Space Agency S.A. (ELDO S.A.).* Published at the Greek Government Gazette B' 882/14.03.2018, p. 10407 et seq.
22. *Explanatory Memorandum to the Draft law on the Authorization of space activities - Registration in the National Registry of Space Objects - Establishment of the Hellenic Space Agency and other provisions.* Accessed from https://www.hellenicparliament.gr/Nomothetiko-Ergo/Anazitisi-Nomothetikou-Ergou?law_id=ca706be4-0ccc-4dbf-a8ce-a83b01095b01 (in Greek).
23. *Law no 3429/2005 titled Public corporations and Agencies.* Published at the Greek Government Gazette A' 314/27.12.2005, p. 5733 et seq.
24. *Law no 2190/1920 on Limited Liability Companies.* Published at the Greek Government Gazette A' 216/19.09.1920, p. 2141 et seq.
25. *Decision of the Minister of Digital Governance ΓΔΟΔΥ/ΔΔΥ/53.* Published at the Greek Government Gazette Υ.Ο.Δ.Δ. 33/18.01.2020, p. 153 et seq.
26. *Law no 5039/2023 titled support measures for the victims' relatives and for the victims of the railway accident of Tempe dated February 28, 2023, pension provision, arrangements for strengthening traffic safety, provisions to enhance development, interventions for the modernization of tourism legislation and other urgent provisions.* Published at the Greek Government Gazette A' 83/03.04.2023, p. 4221. et seq.
27. Hel.A.S. *Greek National Committee for Astronomy [Εθνική Αστρονομική Επιτροπή].* Accessed from https://helas.gr/gnca/index.html
28. *Legislative Decree 3725/3-8-1957.* Published at the Greek Government Gazette A' 144/08.08.1957.
29. *Royal Decree dated 20/06/1958.* Published at the Greek Government Gazette A' 92/20.06.1958.
30. Source. Accessed from https://www.astronomia.gr/wiki/index.php?title=Εθνική Αστρονομική_Επιτροπή
31. Hellenic Astronomical Society, The Constitution of the Hellenic Astronomical Society. Accessed April 2022, from https://helas.gr/const.php
32. Hellenic Astronomical Society, Directory of Hellenic Astronomical Society. Accessed April 2022, from https://helas.gr/members.php
33. Hellenic Astronomical Society. *The establishment of Hel.A.S.* Accessed April 2022, from https://helas.gr/about.php
34. HASI. (2024). *About us, Association's profile.* Accessed July 2024, from https://www.hellenic-asi.org/en/about-us/profile
35. Hellenic-Asi. *Codified amending statute of the Hellenic Association of Space Industry (HASI).* Accessed July 2024, from https://www.hellenic-asi.org/poioi-eimaste/katastatiko (in Greek).
36. Corallia. *Activity fields, si-Cluster.* Accessed July 2024, from https://si-cluster.gr/
37. Corallia. *Activity fields, si-Cluster, Who we are.* Accessed July 2024, from https://si-cluster.gr/about-us/
38. ESA. *Greece creates its first national cluster in the field of space: si-Cluster.* Accessed July 2024, from https://www.esa.int/Space_in_Member_States/Greece/E_Ellhada_demioyrgehi_to_prhoto_ethnikho_cluster_gia_to_dihastema_si-Cluster

Chapter 5
Greek Legislation on Space Activities

Having presented the State and non-State authorities competent to decide on space matters, this chapter will be dedicated to the examination of the substantive law rules applying to space activities conducted under the jurisdiction of the Greek State. Said rules were established in the Greek space legislation which was adopted in 2017 [1],[1] to first complement and clarify the norms established in international space law but also to strengthen the Greek State's role as an emerging space faring country.

– *By way of introduction, the rationale of the 2017 Greek space law*

According to its Explanatory Memorandum [3], a *first rationale* behind the Greek space legislation was to—initially—protect the significant investments made by space operators in infrastructure and technology, in a field that is continually growing in importance. In particular, mention was made (in the final draft of the Greek Space law) to the fact that promoting space activity is a long-term process necessitating legal certainty, as space activities require careful planning and important resources, while involving numerous actors. Parallel to that, it was acknowledged that the space sector has opened up new commercial opportunities in a wide array of applications, making its regulation a priority for the government [3: 26 et seq.]. In this context, the main motivation to enact the national space law was to essentially provide a reliable framework, able to attract and protect private entities—involved in all related fields, such as telecommunications, navigation, national security, public health etc.—operating under Greek jurisdiction.

To elaborate on this view, it was underlined that the mere transposition of international law treaties into the Greek legal order was insufficient; additional

[1] Interestingly, a number of established space-faring countries were likewise late in adopting their space legislation compared to when they first started implementing space activities. For example, France only adopted a comprehensive legal framework in 2018; the same applies to Austria which has traditionally conducted space activities as a member of the ESA, and only adopted its national space law in 2011, to regulate the launch of its two small satellites, TUGSAT-1 and UniBRITE etc. [2: 76 and 86].

A. Koskina, M. Plionis, *Greek Space Law*, Space and Society,
https://doi.org/10.1007/978-3-031-72813-6_5

rules were necessary to complement the general ones, create an attractive framework, provide legal certainty for operators and eventually encourage investments that would strengthen domestic space activity. Hence, although Greece is a relative newcomer in the context of space, the progress achieved in the past decades made it clear that the time had come to adopt a national framework capable of promoting and protecting public and private actors exploring or using space resources—orbital positions and corresponding frequencies—and enabling them to conduct activities in line with the duty of sustainable use, taking into account the rights of other operators and States [3: 27 et seq.].

Strictly speaking, Greece might not be able—at least, in the near future—to invest in expensive infrastructure (like space systems, spaceports or other technologies aimed at launching or receiving spacecrafts). Be that as it may, the decline in space robotics costs and the growing affordability of small satellites (i.e., small-sats) have enabled the development of small-scale space projects. As a result, the country became more involved in space operations based on the use of cost-efficient equipment, such as is the development and deployment of small satellites and small-sat constellations [4, 5]; in consequence, actors investing in the space domain have seen—up to now—a steady rise in profits, which in turn requires adapted policies and effective responses to emerging issues [4]. In particular, information from journalistic sources indicates that the total revenues of the Greek space industry increased from 154.7 million euros in 2018, to 178.6 million in 2019 (a rise of around 15.4%), and then to 211.3 in 2020 (a rise of 18.3%) [4].

As a *second rationale*, it was noted that the increase in commercial (private law) actors conducting space activities in the public interest—such as telecommunications—should by no means result in any loss of control by the Greek State over (said) space activities; the State is, in any case, obliged to adhere to all relevant international space law commitments and comply with all its obligations. Hence, the country needed to ensure that activities conducted by private operators (i.e., for which it is responsible) would be carried out in line with international space law and fulfill its duty of continuous supervision, such as via the clarification of the authorization procedure and the establishment of a national registry for space objects. According to the Explanatory Memorandum of law n° 4508/2017 [3: 26–27], Greece needed a specific legal framework that would be well adapted to space activity, but also a national space authority having explicit competences on related matters. The underlying idea was to adopt a legislation that would make the country's presence felt both at European and international level, while attracting investments—from both public and private actors, contributing to employment and competitiveness in connected sectors of the economy—and ultimately enabling Greece to also play a role in the international arena [3: 28].

Against this background, and in line with the general structure of the Greek space law (namely, law n° 4508/2017 entitled ***Authorization of space activities—Registration in the National Registry of Space Objects—Establishment of the Hellenic Space Agency and other provisions*** [1]), the present chapter will be divided into four sections on: the scope of application of the Greek space law (Sect. 5.1); the rules of substantive law applying to the setting-up and implementation of space activities, in general (Sect. 5.2); the risks and hazards possibly materializing in the context of space operations (Sect. 5.3) and, finally, the oversight and completion of space endeavors (Sect. 5.4).

5.1 Clarifications Regarding the Scope of the Greek Space Law

The four fundamental international space law treaties—namely, the Outer Space Treaty/OST (1967), the Rescue Agreement (1968), the Liability Convention (1972) and the Registration Convention (1975)—establish explicit obligations binding on all States that have ratified them, Greece being one of them. However, some of these obligations are general in their wording, making them not fully clear to the parties and space stakeholders in all their aspects and complexity, while leaving room for different interpretations likely to end up before the courts.

Hence it is underlined, for instance, that "the OST does not define the term 'national activities' or the term 'appropriate use'. This leaves ambiguity with respect to the question of when a State party needs to regulate space activities—is it those activities over which they have jurisdiction under international law, those activities over which they have effective control or any activity in which one of the State's nationals, person or juridical is involved? Furthermore, international space law does not demarcate the boundary between airspace and outer space, nor does it define exactly what activities should be considered space activities. Accordingly, depending on the delimitation applied by a State party, certain activities may be considered space activities, while a different delimitation applied by another State Party results in the same activities not being considered as space activities, for example, suborbital flights" [2: 137–138].

In this context, it soon became apparent that national legislators play an important role in the process of clarifying the rights and obligations—of all actors involved in space activity—under the international space law treaties. Be that as it may, diverging approaches on key issues were brought to light from the outset, such as in the case of defining the clear outward limits of national sovereignty [6]; indeed, following on from that, no international agreement has yet been reached on, amongst other things, the demarcation of outer space from airspace, giving rise to different interpretations and approaches, at the national level.[2]

[2]For instance, contrary to the general trend (to examine all space activities under a common framework) the Explanatory Memorandum to the Dutch Space Activities Act distinguished between "activities in space *stricto sensu* (i.e., activities that take place completely in outer space, such as manned space activities); activities in space *largo sensu* (i.e., activities that are conducted on Earth but the consequences occur in outer space, such as the operation and guidance of space objects) and space activities *largo sensu* (i.e., activities in which both the activities themselves and the consequences do not exclusively occur in outer space, such as the launching of space objects)" [2: 69].

Thus, as part of the general effort to promote harmony between national regulations and international space law, the Sofia Guidelines made it clear that the initial provisions of domestic space legislations should aim *first* at clarifying their scope of application, to minimize any risk of conflict of laws as to the regulation of space activities. In other words, the adoption of an objective and genuine link between a space activity and a State was regarded as being of major importance for the proper application of national laws [7: Art. 1]. Based on this reasoning, the Greek space law clarified its territorial and personal scope of application, subsequently its material scope of application and, in addition to that, a few complementary key concepts.

5.1.1 Territorial and Personal Scope of the Greek Space Law (i.e., Scope rationae loci *and* rationae personae*)*

National legislators are expected to initially define the territorial and material scope of their domestic laws. Therefore, in theory, (i) the *first aim* of the Greek legislator was to define or enumerate the activities and persons subject to Greek law, and set out preconditions for the exercise of national jurisdiction; and (ii) *second*, to (ideally, also) establish a clear position on the delimitation of outer space, where the core of said space activities are to primarily take place, in order to enhance the rational governance of space activities under Greek jurisdiction.

In this vein, the Greek space law opens with an introductory provision, aimed at defining both its scope of application and the general contents of the law. More precisely, Art. 1.1.(α) of law n° 4508/2017 [1] provides that the law shall regulate the terms and procedures for authorization; liability for damage caused by space objects; and the supervision of space activities. Then, it mentions that the law shall also provide for the setting up and maintenance of the National Register for the recording and registration of space objects [1: Art. 1.1.(β)].

Following this introductory provision, the scope of the Greek space law is further explained, largely in line with the Sofia Guidelines which make reference to an "effective link (that must exist between the space activity and) the country enacting the specific law", justifying the application of the domestic rules in question [7: Art. 1—comment].

5.1.1.1 Establishment of Both a (Quasi)territorial and Personal Jurisdiction

More precisely, regarding (i) the preconditions for exercising jurisdiction, the Greek law lays down that it shall apply to space activities that are carried out *within the Greek territory* [1: Art. 1.2.(α)] or *outside it* provided that, in this second case, the infrastructure, personnel, and/or the movable and immovable property which is being used belongs to the Greek State or falls under its jurisdiction [1: Art. 1.2.(β)]. Additionally, the law may apply to space activities regardless of their location, but only in the event that such activities are carried out by natural persons possessing Greek nationality or—alternatively—by legal entities established within the Greek territory, under the specific condition that such thing is provided for by an international agreement or treaty [1: Art. 1.2.(γ)]. Finally, the Greek space law shall also apply in those cases where activities are carried out by (domestic and/or foreign) natural or legal persons within the Greek territory, in case infrastructure, personnel, and/or movable or immovable property of the Greek State is being used; or if they—i.e., the natural or legal persons—fall under the Greek State's jurisdiction and are placed under its responsibility as the "launching State" in accordance with International Law [1: Art. 1.2.(δ)].

Hence, from a legal perspective, the Greek legislator established—*first of all*—both a territorial[3] and a personal[4] jurisdiction, similarly to other national legislators.[5] Interestingly, though, the Greek law did not reiterate the specific wording of Art. VIII of the OST on the quasi-territorial jurisdiction of States, i.e. allowing States to retain jurisdiction over their space objects while they are in space or on celestial bodies (e.g. no reference is made to the applicability of Greek law *precisely* to space objects when they are operating in space). However, the Greek space law explicitly mentions the use of movable and/or immovable property belonging to the Greek State or falling under its jurisdiction [1: Art. 1.2.(β) and Art. 1.2.(γ)] *lato sensu*, as an element to take into account, which in fact amounts to the same thing: it likewise establishes the quasi-territorial jurisdiction of the Greek State (i.e., over its operating space objects as well).

Essentially, all the criteria that were laid down for the application of the Greek space law—namely, the location of the space activity (i.e., "within the Greek territory"); the ownership of movable or immovable property being used for this purpose (i.e., "belonging to the Greek State or falling under its jurisdiction") and/or the nationality of the natural or legal persons involved in the space-related activity (i.e., "natural persons possessing Greek nationality")—seem to reflect the idea of an effective link and genuine concern of the Greek State. From this perspective, the Greek space legislation appears well-suited to the purpose of avoiding the emergence of any 'flags of convenience' problems.

This conclusion is reinforced by the fact that Art. 1.3 stipulates that the Greek State shall be held liable for space activity, in line with the obligations undertaken under the international space law treaties.[6] The inclusion of this rule in Art. 1 of law

[3] "The so-called 'territorial jurisdiction', the right of a sovereign state to rule over activities conducted on or from its territory, presents the other universally-recognized basis for exercising jurisdiction" [8].

[4] "The nationality of natural or juridical persons undertaking certain activities represents one of the universally-recognized bases for the exercise of jurisdiction" [8].

[5] Such as e.g. in Sweden, Australia, China, France, Austria, Indonesia, Denmark and New Zealand [2: 140–141].

[6] Interestingly, law n° 4508/2017, Art. 1.3 does not specify which particular treaties of international space law should be taken into account, which therefore leaves a certain margin of discretion to the competent authorities.

n° 4508/2017 and under the title *Subject matter—Scope* serves to also make it clear and explicit, right from the outset, that the Greek State shall assume full responsibility for all its space operations.

Finally, to avoid any (jurisdictional) problems that could arise from the registration—or non-registration—of space objects, it is clarified that, *first*, in cases where other States meet the requirements as launching States pursuant to Art. II para. 2 of the Registration Convention, the competent Minister—i.e., today, the Minister of Digital Governance—is the authority entitled to decide on the registration of a space object in the *Greek National Register for the recording and the registration of space objects* [1: Art. 1.4]. *Secondly*, Art. 1.5 provides *expressis verbis* that the provisions of law n° 4508/2017 will not apply to any space activity related to or using space objects which are listed in the National Register for space objects of another country and thereby falls under that country's international responsibility for damage, in line with international law.[7]

5.1.1.2 The Greek Position on the Delimitation of Outer Space

Regarding the second point—namely, (ii) the official position that a State may have on the definition and delimitation of outer space—, the Greek space law is silent, as officially communicated to the UN COPUOS LSC in 2021: "Given the current level of space and aeronautics activities in Greece, national legislation defining or delimiting outer space has yet to be put in place" [9].

However, taking into account the two principal approaches to this specific challenge (i.e., the spatialist and the functionalist approach), the Greek State has established that: "Not only the differentiated capacity among States to exercise their sovereignty in any part of space, but also the prohibition of national appropriation by claim of sovereignty or by means of use or occupation, as confirmed by Art. II of the Outer Space Treaty (1967), would undeniably pose critical problems in a possible adoption of a spatial theory. Additionally, should the boundary be based in relation to anti-satellite tests or the activities of certain States to remove and destroy own satellites, it would be at least at an altitude equal to that of satellite orbits. This solution would not serve as a clear boundary between airspace and outer space.

In contrast, it is recommendable to promote a functional approach, given the current and foreseeable technological state. According to this theory, space is to be considered outer space at any distance from the surface of the Earth as long as it may be used by space objects, i.e., capable of performing space flight. Namely, the different nature of space activities, and the fact that there is no connection with the underlying territory, implies that these activities shall, wherever conducted, be exclusively subject to the sovereignty of the launching States. Therefore, the legal

[7] In such a case, the space operator must inform the competent Minister (i.e., the Minister of Digital Governance) about the space activity in question, without any delay [1: Art. 1.5].

regime of outer space should be determined based on the capacity of space launches or orbits of space devices, at their lowest perigee" [10].[8]

5.1.2 *Material Scope of the Greek Space Law* (Scope rationae materiae)

In conjunction with the definition of its territorial and personal scope (*rationae loci* and *rationae personae*), law n° 4508/2017 enumerates a number of activities which fall under the Greek space law, due to their particular features and characteristics (*rationae materiae*). A comparative analysis of national space laws shows that "(i)n essence, and although different terms are used in the various acts, the national space legislation of (. . .) States regulates the launching, operation, control and return of space objects" [2: 142][9]. Hence, the Greek legislator followed the same pattern, but additionally defined a few related terms with the intent to facilitate the reading and application of the law.

Before entering into the substance, it is crucial to mention that, *ab initio*, States are free to conduct *any* space activity they deem necessary and appropriate, in line with their own interests, except for those that are explicitly prohibited by international space law. Precisely, Art. I para 2 of the OST stipulates that "Outer space, including the Moon and other celestial bodies, shall be free for exploration and use by all States without discrimination of any kind, on a basis of equality and in accordance with international law". Read in conjunction with Art. VI et seq. of the OST, the above essentially means that—as a general rule—, all States may carry out activities in outer space, provided they guarantee that these are conducted taking into account the interests of other States and in conformity with the provisions set forth in the treaty.[10] In other words, countries must refrain from carrying out space operations that are explicitly banned or restricted by international space law. The aim is to ensure the peaceful and sustainable use of space resources to the benefit of all, and avoid the creation of (additional) risks for the conduct of space operations; a typical example of such a prohibition being the ban on activities aimed at placing "in orbit around the Earth any objects carrying nuclear weapons or any other kinds of weapons of mass destruction, install such weapons on celestial bodies, or station

[8] This document reaffirmed the Greek position as communicated in 2017 [11].

[9] For the sake of precision, it is necessary to mention that the study refers to the national space laws and regulatory frameworks of Sweden, the UK, Belgium, the Netherlands, France, Austria and Denmark.

[10] Precisely, Art. VI of the OST lays down that: "States Parties to the Treaty shall bear international responsibility for national activities in outer space (. . .) whether such activities are carried on by governmental agencies or by non-governmental entities, and for assuring that national activities are carried out in conformity with the provisions set forth in the present Treaty. The activities of non-governmental entities in outer space, (. . .) shall require authorization and continuing supervision by the appropriate State Party to the Treaty (. . .)".

such weapons in outer space in any other manner" [12: Art. IV para. 1]. Therefore, as long as they fully comply with the rules of international space law, States are free to conduct space activities and, at the same time, adopt national space legislation to regulate these activities in more detail.

Against this background, the Greek space law [1], having established that it applies to space activities reflecting an effective link with the Greek State (as explained above), also defines the term *space activity*, to clarify which precise operations are being regulated. Hence, Art. 2.6 of law n° 4508/2017 adopts a descriptive approach to space activity, stipulating that such activity may consist in the launching, flight-operation and maneuvering activities, as well as in activities related to the return of space objects. In light of this wording, it may be noted that the precise concept of space activity is therefore defined on the basis of operational criteria, rather than any (scientific or commercial) explicit purpose.

To mention but a few examples of different approaches adopted in national laws as points of reference, the Indonesia Space Act takes into account the key purpose of space activities, such as by making explicit reference to space science, remote sensing and commercial space activities;[11] correspondingly, the Danish Outer Space Act also includes *other essential activities in this connection*, although it does not further define this very particular expression.[12]

Contrary to the above, the Greek legislator mainly reproduced the approach adopted in the Sofia Guidelines. Indeed, the latter document stipulates that the concept of space activity may refer to "the launch, operation, guidance and re-entry of space objects into, in and from outer space and other activities essential for the launch, operation, guidance and re-entry of space objects into, in and from outer space" [7: Art. 2], which seems to have been the basis for the wording of Art. 2 of law n° 4508/2017. This observation is corroborated by a specific mention—in the Explanatory Memorandum of the Greek space law—that, in formulating Art. 2, the national legislator gave consideration to the definitions set forth in international space law instruments [3: 2].

In the same vein, the Greek space law also clarified the concepts of *flight operation and guidance* (being mentioned in Art. 2.6), in its Art. 2.5. Effectively, it lays down that: 'flight operation and guidance' may refer to manoeuvers concerning the placement in orbit; the flight conditions; or the navigation, trajectory or launching of a space object such as, mainly, by choosing, reviewing or correcting its course or trajectory (a few other national space legislations have also explained

[11] "The activities that are explicitly considered to be space activities are space science, remote sensing, space technology capability, launching and commercial space activities", the latter being defined as "any activity that aims to gain an economic return" [2: 97].

[12] Nonetheless, "(i)t has been clarified that activities such as the participation of Danish researchers or students in satellite projects of foreign space agencies or companies do not fall within the definition" [2: 104]. .

terms in a comparable manner;[13] such as e.g., the Belgian law[14] or the Dutch law[15]).

For the sake of completeness, mention can finally be made of particular proposals—that were put forward during the public consultation on the draft of the Greek space law, which took place as part of the regulatory process prior to its adoption [13][16]—, to clearly stipulate that the concept of space activity may also refer to operations of a non-lucrative nature, such as those aimed at scientific, research or educational purposes. The idea behind this proposal was to promote a well-defined (i.e., more lenient) regime for such activities, allowing for derogations and a less costly procedure. On this basis, it was also suggested that such operations should benefit, as a general principle, from simplified authorization (or other) rules [13: see under Art. 3 and Art. 5].

The Greek legislator did not fully adopt the idea of specifically regulating this type of space activities, or establishing a derogatory regime applying to it, at least as a guiding norm from the outset. However, the Greek space legislation still laid down a few derogative substantive rules, allowing to take into account the ultimate goals of the space activities, as well as the means and funds used to achieve them. In this sense, particular conditions that are inherent to space research or educational operations—or to any comparable activities—may justify a differential treatment subject to the discretion of the competent authority.

Hence, an activity will fall under the provisions of the Greek space law (that *prima facie* applies to *all kinds* of space activities) provided it simply falls into the category of "launching, flight-related and maneuvering activities, as well as those related to the return of space objects".

Therefore, the definition of a space activity chosen by the Greek legislator may, on the one hand, be regarded as conventional, as it mainly reiterates the definitions adopted by a majority of States, or the one which is proposed in the Sofia Guidelines. On the other hand, it is also a definition which is flexible and highly adaptable—without however being overly broad or vague—, as it may apply to activities with different objectives, without requiring significant modifications to keep pace with advances in space technology and/or changes in the nature of space activities.

---

[13] However, not the Sofia Guidelines.

[14] In the Belgian law, "(b)oth the terms 'flight operation' and 'guidance' have been clarified as any operation with respect to positioning, flight conditions, navigation or progress of the space object in outer space, such as the choice, control or adjustment of the orbit or trajectory" [2: 57].

[15] The commentary to the Dutch Space Activities Act clarifies that: "flight operations are all operations of the space object during the flight phase, meaning the phase between launch and the final position of the space object in orbit. The guidance of space objects is understood to be activities to control the space object, such as maneuvering, watching out for space debris and tending the fuel levels" [2: 67].

[16] The public consultation procedure, which takes place prior to the adoption of laws, was established in 2012, by law n° 4048/2012 [14]. Therefore, at the time law n° 4508/2017 was being discussed, law n° 4048/2012 was in force, stipulating that comments received during consultation periods *could* be taken into account. All the comments to the draft law n° 4048/2012 remain available on the website devoted to its consultation [13].

5.1.3 *Definitions of Essential Concepts in Law n° 4508/2017*

Complementing all the above, the Greek space law thereafter defined a number of basic and relevant concepts to facilitate its interpretation and application. By way of illustration, taking into account that space objects—and space infrastructure and/or space technology likewise—play a key role in the conduct of space activities, the very concept of *space object* is legally explained. It is, however, interesting to note that Art. 2 of the Greek space law does not simply reiterate the content of the corresponding draft provision proposed in the Sofia Guidelines; it also defines additional terms (such as the concept of space operator) which are in truth necessary to make the Greek space legislation more comprehensible in its entirety.

5.1.3.1 Significance of the Term *Space Object* in the Greek Space Law

As might be expected, Art. 2 of the Greek space law focuses first on one of the most fundamental concepts (i.e., the space object) and provides, initially, three alternative definitions of it. Therefore, according to law n° 4508/2017, a space object may consist of: (i) any object *launched or intended to be launched into orbit around the Earth or to a destination beyond Earth orbit*, in outer space [1: Art. 2.1.(α)]; (ii) any device or system intended *to be used for launching* an object, as referred to in point (α), even if such a device is not carrying any payload, but operating with the sole aim of developing and validating its proper functioning [1: Art. 2.1.(β)] and (iii) *any removable part* of an object, as referred to in points (a) and (b) [1: Art. 2.1.(γ)].[17]

In essence, Art. 2, para. 1 of law n° 4508/2017 reiterates the definition of a space object, as established in international space law *lato sensu*. Indeed, it reproduces *first* the content of Art. 2 of the Sofia Guidelines, defining space objects as "any object launched or intended to be launched into outer space including its component parts as well as its launch vehicle and parts thereof" [7: Art. 2]. As a result of that, Art. 2 para. 1 of law n° 4508/2017 is also in line with the approach established in the Registration Convention and in the Liability Convention, since both Art. I (b) of the Registration Convention and Art. I (d) of the Liability Convention mention that "(t)he term 'space object' includes component parts of a space object as well as its launch vehicle and parts thereof".

Theoretically, a different definition could have been enacted. By way of illustration, the Chinese space law adopts a much more detailed description, namely: "the term 'space object' includes an artificial satellite, crewed spacecraft, space probe, space station, launch vehicle, and parts thereof, and other human-made objects launched in outer space but excludes sounding rockets and ballistic missiles that temporarily cross outer space" [2: 49]. Be that as it may, the significance of each definition must primarily be considered in the context of States' responsibility for space activities conducted according to their capabilities and under their jurisdiction.

[17] Emphasis added.

From this perspective, it is reasonable to expect that the definitions adopted in national space laws will basically reflect—with the aim to regulate—the type of space objects that are usually launched under the jurisdiction of each State concerned. In this sense, it may be argued that the definition of space objects in law n° 4508/2017 is well adapted to the current features (and means of action) of the Greek space sector.

At the same time, it is noteworthy that according to the definitions contained in Art. 2, the Greek law only applies to objects programmed (or launched with the aim) to be placed in Earth orbit or to a destination beyond Earth orbit in outer space, that is to activities referred to as orbital flights.[18] Following this reasoning, other types of activities—such as e.g., suborbital flights[19] which, from the perspective of the Greek State, must *first* be addressed in international space law and are, therefore, considered difficult for the national legislator to regulate through detailed or implementing rules[20]—could be regarded as falling outside the scope of said provision. In truth, although a few countries have already adopted laws to regulate suborbital flights (such as the UK[21]), Greece has made it clear that the regime possibly applying to such flights necessitates a consensus-based delimitation of outer space (i.e., which in turn requires to first agree on either a functionalist or a spatialist approach on this issue).

On this latter point—as already referred to—Greece contends that "the functionalist approach to the definition and delimitation of outer space may provide greater clarity than the spatialist one, as it would result in the application of only one legal regime (i.e., space law) to suborbital flights, whose primary function is to enter outer space (despite the fact that they are also designed to traverse airspace). The functionalist approach distinguishes between aeronautical and astronautical activities on the basis of the primary purpose of a vehicle's design rather than on the location of the vehicle at any single point in time. Under this approach, a company's suborbital operations would be governed under space law only (with the exception of air traffic management rules for that portion of the flight that crosses airspace), because suborbital flights are designed for space travel" [17].

[18] Orbital flights are "characterized by a space object escaping Earth's gravity and reaching a speed great enough to ensure that it "falls past" Earth as it rotates around the planet [15 and references therein].

[19] A suborbital flight is one where "a space object escapes Earth's gravity but does not reach a sufficient speed to outrun the planet and therefore re-enters the atmosphere before it has made a complete rotation around Earth" [15 and references therein].

[20] "Given their unique technological design ... suborbital flights evade regulation under a specific regime due to overlapping subsets of international law. Suborbital vehicles spend time in airspace (requiring the application of air law) as well as outer space (requiring the application of space law)" [15: 388].

[21] According to the UK Space Industry Act adopted in 2018, space activities that fall under its scope may consist in the launch, or procurement of a UK launch "space *or suborbital*, and return", (emphasis added) [16].

As a result, regarding the regulation of suborbital flights precisely, the Greek State made it clear that "(t)he choice whether or not to apply international or national law [*i.e., to suborbital flights*] is related to the content of the definition of suborbital flights and to the specific characteristics of the flight" [17], and that "(t)he definition of suborbital flights for scientific missions or crewed space flights has particularities, making it a complex subject that requires careful study, which we believe should be done within the competent United Nations bodies, following a wide-ranging, open consultation" [17]. Hence, following this line of reasoning, the Greek space law was adopted to primarily apply to objects launched to be placed in Earth orbit or sent to a destination beyond Earth orbit in outer space, and to any objects facilitating or used in this purpose. In reality, the importance of being placed in orbit as a key characteristic of space activities (and thus as a precondition for the application of the Greek space legislation adopted in 2017) also lies in the explicit wording of Art. 2 para. 1.(a),[22] the reaffirmation of this specific prerequisite in Art. 2 para. 1.(β),[23] as well as in the references to objects deployed in orbit found in Art. 2 para. 2[24] and Art. 2 para. 5.[25]

Finally, the Greek legislator went a step further by adding—in Art. 2 para. 1.(β)—a few additional technical explanations. Indeed, said article mentions that the concept of space object may also refer to any device or system intended to be used for launching an object, as described in Art. 2 para. 1.(α), "even in the case such a device is not carrying any payload, operating with the sole aim of developing and validating its own functioning". Nevertheless, it is at the same time noteworthy that specific concepts used in these clarifications (like the term *payload*) are not legally explained or defined, contrary to what is established in other national space legislations.[26] As a result, these terms should arguably be interpreted having regard to their purpose and, in any case, also considering the perspective of space scientists. This interpretative approach could apply *mutatis mutandis* to another general

---

[22]Law n° 4508/2017, Art. 2 para. 1.a: "any object launched or intended to be *launched into orbit around the Earth* or to a destination *beyond Earth orbit, in the Outer Space*", (emphasis added).

[23]Law n° 4508/2017, Art. 2 para. 1.β: "any device or system intended to be used for launching an object, *as referred to in point (a)*", (emphasis added).

[24]Law n° 4508/2017, Art. 2 para. 2, last sentence: "If a space object cannot be maneuvered *after being put in orbit*, the operator shall be the person giving the order to *place the space object in orbit*", (emphasis added).

[25]Law n° 4508/2017, Art. 2 para. 5: "flight operation and guidance: any procedure concerning the *placement in orbit*, flight conditions, navigation, trajectory, or *launching of a space object in outer space* such as, mainly, the choice, review, or *correction of its course or orbit trajectory*", (emphasis added).

[26]By way of comparison, the New Zealand Space Activities Act clearly defines *first* the launch vehicle (i.e., a launch vehicle or any component part of a vehicle that reaches, or is intended to reach, outer space on its own or carries or supports the launch, or is intended to carry or support the launch of, a payload) *but also the payload*: "a payload is the object that is carried or placed, or is intended to be carried or placed, in outer space, including component parts of the launch vehicle that are specifically designed for the object, and a load to be carried for testing or non-profit purposes" [2: 112].

concept—mentioned in Art. 2 para. 1.(γ)—namely, to *any structural component*[27] of the objects defined in para. 1.(α) and 1.(β) of Art. 2.

5.1.3.2 The Key Persons Involved: Space Operators and Manufacturers

Further to the above, it is the concept of *space activity operator* (or operator) which is thoroughly defined. More precisely, in the Greek space legislation, the operator is defined as any natural or legal person carrying out—or undertaking to carry out—a space activity. As a result, the operator is directly responsible for ensuring, either individually or together with any other operator(s), *the effective control* of the space object in accordance with explicit terms that have been established in a related agreement or authorization (Art. 2.2).[28] For the sake of precision, the same article (a few lines further on, under Art. 2.3) defines the term *effective control* as the activation of any control and command system—or of any remote control and command system—, as well as/or of the monitoring system that is necessary to carry out the launching, *flight operation and guidance* of one or more space objects (this latter expression is also specifically defined in the law).

Notably, to ensure clarity, law n° 4508/2017 also provides a detailed explanation of the concept of *flight operation and guidance*: such activity may refer to any operation aimed at the placement in orbit, and/or at ensuring the flight conditions, navigation, trajectory, and/or the launching of an object in outer space; it could also refer—amongst others—to the choice, review, or correction of its course or orbit trajectory.[29] In fact, this definition (i.e., the definition of the concept of flight operation and guidance) is crucial because it is used, *first*, to define space activities,[30] but also to establish the terms of application of Art. 3 on the conditions for authorizing a space activity, or of Art. 9 on the revocation of such authorization.

As a guiding principle, law n° 4508/2017 established the fundamental rule that the space operator is the person or entity responsible for carrying out the space activity.[31] Therefore, the designation of the operator is of paramount importance and, in cases where more than one operator is involved, requires eliminating any doubts or ambiguity about their identity, or the specific space activities they may initiate. This practical reasoning is apparently the rationale underlying the provision

[27] Emphasis added.

[28] Emphasis added. According to Law n° 4508/2017, Art. 2.2, in the event that a space object cannot be maneuvered after being placed in orbit, the operator shall be considered the person who gave the order to place this space object in orbit.

[29] Law n° 4508/2017, Art. 2 para. 5: "flight operation and guidance: any procedure concerning the *placement in orbit*, flight conditions, navigation, trajectory, or *launching of a space object in Outer Space* such as, mainly, the choice, review, or *correction of its course or orbit trajectory*", (emphasis added).

[30] See, law n° 4508/2017, Art. 2.6.

[31] See, for example, law n° 4508/2017, Art. 11 "International liability for damage".

of Art. 2.2 which explains in such detail what is meant by carrying out—or undertaking to carry out—a space activity.

At the same time, the Greek space legislation defines the concept of *manufacturer* of a space object. Literally, the manufacturer is the person who participates or has participated in the development, manufacture or assembly of all or part of a space object [1: Art. 2.4]. Such a reference—i.e., to the manufacturer of a space object—is in reality explained by the fact that the manufacturer and the operator may not always be the same natural or juridical person.[32] Put differently, the manufacturer creates the space object, but does not (necessarily) operate it. In this regard, it is important to note that the description of the tasks to be performed by a manufacturer (namely, the development, manufacture or assembly) is quite general; hence, it remains to be seen which technical activities really fit into this definition, considering aspects such as space industry trends and practices.

Be that as it may, by defining the manufacturer (also) as a person participating in the *assembly of all or part of a space object*, Art. 2.4 of law n° 4508/2017 most likely refers to the emerging practice adopted by both manufacturers and/or space operators, of buying space objects components with the aim of having them assembled afterwards. Indeed, the satellite and launching vehicles industry improves, and the demand grows for components of space structures, electronics and supporting equipment; as a result, space objects components are developed by technology providers, sold by vendors and merely assembled by the final consumer (e.g., in this case, the space manufacturer) [18, 19].[33]

Nevertheless, it is also noteworthy that the Greek space law does not make any other reference to the manufacturer of a space object, except in Art. 17 enumerating those pieces of information that must be attached to a request for authorization of a space activity [1: 17.3.(α)]. From this perspective, it appears that the Greek legislator's interest in the manufacturer of a space object is principally to ensure that any information related to the launching of a space object—*lato sensu*, namely starting from its development and including all the persons involved in all phases—is effectively brought to the attention of the competent national space authorities.

Finally, one may note that, in a rather different direction than mentioned above, the Sofia Guidelines' definition of the operator is more general and broader, as it is intended to accommodate the various situations which may arise in the space sector across different national contexts; precisely, the Guidelines establish that the term simply refers to "a natural or legal person carrying out space activities", while making no mention of the word or concept of manufacturer. At the same time, the instrument explicitly underlines that the enumeration in draft Art. 2, entitled *Definitions-Use of Terms*, "is by no means exhaustive. National legislators are free to add

[32] The same comment was made in reference to the *owner* defined in the Danish space legislation [2: 105–106].

[33] *Inter alia*: "(. . .) a specialty space data partner that helps a company build an Earth observation platform from soup to nuts. That means designing the sensors, the bus, the navigation and telemetry systems, and the downlink and ground infrastructure needed to support this data-heavy industry." [19].

more definitions if they so consider", eventually considering the specific features of their own space sector; which, in practice, the Greek legislator did.[34]

5.1.3.3 The Concepts of Launching State, State of Registry and Other Clarifications

The concepts of *Launching State* and *State of Registry* are not specifically defined in the Sofia Guidelines. However, these topics are covered by the relevant international space law treaties and by the 2004 Resolution on the Application of the Concept of the 'Launching State' [21], as well as by the 2007 Resolution on the Recommendations on Enhancing the Practice of States and International Intergovernmental Organizations in Registering Space Objects [22].

Regarding, first of all, (i) the concept of launching State, mention must be made to the fact that its definition was initially established in Art. VII of the OST. According to this provision, "(e)ach State (. . .) that *launches or procures the launching* of an object into outer space (. . .) and each State Party *from whose territory or facility* an object is launched, is internationally liable for damage to another State Party to the Treaty or to its natural or juridical persons (. . .)[35]". Building on this legal foundation, Art. 1 (c) of the Liability Convention and Art. I(a) of the Registration Convention clarified that: "(t)he term 'launching State' means: (i) A State which launches or procures the launching of a space object; (ii) A State from whose territory or facility a space object is launched". Hence, the definitions converge.

As a result, law n° 4508/2017 adopted a similar wording. Indeed, Art. 2.11 reads that "the Launching State is defined as the State launching or procuring the launch of a space object" [1: Art. 2.11.(α)], or "the State from whose territory or infrastructure a space object is launched" [1: Art. 2.11.(β)]. Importantly, Art. 2.11.(β) of law n° 4508/2017 also mentions that these definitions must be interpreted in conjunction with those set out in Art. VII of the Outer Space Treaty, Art. I of the Liability Convention and Art. I of the Registration Convention. While this clarification might initially appear redundant, it is likely that the Greek legislator essentially aimed to also confirm, in this way, adherence to any acceptable future interpretations of the definitions established in the international space law treaties which are explicitly referred to.[36]

---

[34]One could postulate that, as the Greek space sector has only recently started to develop a strong presence in the international arena (see, *inter alia* [20]), it is likely that—until now—the domestic space industry has largely relied on importing space objects components; which could also explain the specific inclusion and definition of the term *manufacturer* in the Greek space law.

[35]Emphasis added.

[36]Law n° 4508/2017, Article 2 reads: "For the purposes of this Law the following definitions shall apply: (. . .) 11: "Launching State": (α) a state launching or procuring the launch of a space object, (β) a state from whose territory or facility a space object is launched, according to the definition provided in Article VII of the Outer Space Treaty, in the first Article of the Liability Convention or in the first Article of the Registration Convention".

Contrary to that, in defining (ii) the State of Registry, the Greek space law makes only a brief—though substantial—mention to Art. I of the Registration Convention which, in any case, remains the primary point of reference for this notion. As a matter of fact, the OST only implies the concept of State of Registry without assigning it a very precise meaning. In particular, the term is referred to *first* in Art. V para. 1 of the OST on the assistance to, and the return of, astronauts, laying down that "When astronauts make such a landing, they shall be safely and promptly returned to the *State of registry* of their space vehicle[37]"; likewise, Art. VIII of the OST—on the jurisdiction and control over space objects—only states that: "A State Party to the Treaty *on whose registry an object* launched into outer space *is carried* shall retain jurisdiction and control over such object, and over any personnel thereof, while in outer space or on a celestial body".[38]

Consequently, to elaborate on this fundamental concept, Art. I(c) of the Registration Convention established that "(t)he term 'State of registry' means a launching State on whose registry a space object is carried in accordance with Article II"; following on from this, Art. II of the Registration Convention stipulates the cases in which a space object must be recorded in a National Registry.[39] In this context, Art. 2.14 of law n° 4508/2017 simply laid down that "the State of Registry is a launching State on whose registry a space object is listed, in accordance with the first Article of the Registration Convention". Here again, it seems that, by defining the concept of State of Registry through a specific reference to international space law, Art. 2.14 of law n° 4508/2017 primarily aimed to incorporate into the national space law framework any agreed interpretations of the Registration Convention, particularly regarding the articles and concepts it refers to.

The instruments of international space law mentioned above aim at mainly ensuring that States have full knowledge and control over their space activities, and that they implement all their duties and obligations under the said framework.[40]

---

[37] Emphasis added.

[38] Emphasis added; Art. VIII of the OST mentions the concept of State of Registry a second time at the end of said article without, again, defining it: "Such objects or component parts found beyond the limits *of the State Party to the Treaty on whose registry they are carried* shall be returned to that State Party, which shall, upon request, furnish identifying data prior to their return", (emphasis added).

[39] Art. II para. 1 of the Registration Convention reads: "When a space object is launched into Earth orbit or beyond, *the launching State shall register the space object* by means of an entry in an appropriate registry which it shall maintain (. . .)", and Art. II para. 2 clarifies that "Where there are two or more launching States in respect of any such space object, *they shall jointly determine which one of them shall register the object* in accordance with paragraph 1 of this article (...)", (emphasis added).

[40] "The international legal status of a launching State is determined through the application of appropriate treaties (e.g. the Outer Space Treaty, the Registration Convention, and the Liability Convention) and cannot be renounced unilaterally because of international obligations that flow from such a status. Unilateral renunciation of a State's international obligation might be considered "an internationally wrongful act" on the part of that State, which is obliged to fulfill its international obligations in good faith" [23].

In this context, and bearing in mind the importance of providing as much legal clarity as possible—to facilitate the safe conduct of space activities—the Greek legislator also defined other concepts such as the *transfer of authorized space activities* (which signifies the transfer of an authorization and/or the transfer of rights related to space activities, in general) [1: Art. 2.13], and the concept of *damage*.

Contrary to other concepts, the term *damage* is extensively addressed *firstly* in the OST, Art. VII, which stipulates that "Each State Party to the Treaty that launches or procures the launching of an object into outer space (. . .) and each State Party from whose territory or facility an object is launched, is internationally liable for damage to another State Party to the Treaty or to its natural or juridical persons by such object or its component parts". *Secondly*, it is addressed in the Liability Convention, as the damage that may arise in the context of space activities constitutes the core subject matter of this particular treaty, which focuses on defining and clarifying the meaning and scope of the rule established in Art. VII of the OST.

Thus, in dealing with the issue of damage caused by a space activity, which is both complex and delicate, the Greek legislator adopted a rather cautious approach by simply reiterating the Liability Convention provisions. More precisely, law n° 4508/2017 only stipulated that the term *damage* refers to any type of damage "as it is defined in the first article of the Liability Convention" [1: Art. 2.12]. In this manner, the national space law essentially created a flexible and future-proof framework, allowing—again—for the incorporation into domestic law of any developments in the definition and interpretation of damage caused in the context of space activity.

Following from the above, one could argue that, overall, the national legislator did not merely transpose the full text of the Sofia Guidelines draft articles into the Greek legal order, as no definitions of the terms of *authorization*; *supervision* and *commercial space activity* mentioned in said Guidelines [7: Art. 2] were included in law n° 4508/2017. On the other hand, other concepts—such as, *effective control*; *flight operation and guidance*; *damage* and *transfer of authorized space activities*—were introduced and defined with precision, as the domestic legislator most probably took the specificities of the Greek space sector into account, as much as possible.

A final consideration with respect to the definitions (and scope of application) that were established in Art. 2 of the Greek space law is that said provision eventually makes explicit reference to the four (4) principal international space law treaties which have been integrated into domestic law—namely the Outer Space Treaty or OST; the Liability Convention; the Registration Convention and the Rescue Agreement (mentioned in this specific order, in the Greek space law)—, as well as to their instruments of ratification.[41] In this manner, the legislator indirectly pointed out that all the international space law treaties ratified by Greece are integral parts of national law and prevail over any contrary provision of the law, as it is in any case clearly established in Art. 28 para. 1 of the Greek Constitution.

Having defined the basic terms and concepts of the current regulatory framework establishing its scope of application, law n° 4508/2017 then determines the

[41] See, for more details, Sect. 2.1.

substantives rules of law that apply to space activities conducted under the jurisdiction of the Greek State, in a manner that complements European and international space law.

5.2 Setting-Up and Implementation of the Space Activity

International space law treaties—and mainly the Outer Space Treaty/OST, which is regarded as the cornerstone of international space law—established the fundamental right to explore and use outer space for peaceful purposes, to the benefit of all countries. As a result, States may set up and implement a wide array of space activities, provided they are conducted in accordance with international space law, including related declarations, principles and guidelines contributing to its interpretation.

More precisely, the OST—signed in 1967, ratified by Greece and incorporated into the national legal order by means of legislative decree n° 670/1970—established a specific legal framework essentially aimed at ensuring the freedom to explore space and use its resources, taking into account the interests of all actors involved. Indeed, both Art. I and Art. III of the OST stipulate that States are free to explore and use outer space, in line with international law. Moreover, Art. III clarifies that State parties to the treaty shall conduct space activities "in the interest of *maintaining international peace and security and promoting international cooperation and understanding*[42]".

The freedom to explore and use outer space is subject only to those limitations that are clearly set out in the OST and other space law treaties. Hence, by way of illustration, the OST stipulates—in Art. VI—that: "States Parties to the Treaty shall bear international responsibility for national activities in outer space, including the Moon and other celestial bodies, whether such activities are carried on by governmental agencies or by non-governmental entities, and for assuring that national activities are carried out in conformity with the provisions set forth in the present Treaty. The activities of non-governmental entities in outer space, including the Moon and other celestial bodies, *shall require authorization and continuing supervision* by the appropriate State Party to the Treaty (...)".[43] In addition to that, Art. VII of the OST laid down rules on the liability of States for damage arising from the conduct of space activities, whether towards other States or to (their) nationals and/or property[44] (*n.b.*, a substantial body of research has been devoted to gaining a better

[42] Emphasis added.

[43] Emphasis added.

[44] OST, Art. VII: "Each State Party to the Treaty that launches or procures the launching of an object into outer space, including the Moon and other celestial bodies, and each State Party from whose territory or facility an object is launched, is internationally liable for damage to another State Party to the Treaty or to its natural or juridical persons *by such object or its component parts on the Earth, in air space or in outer space, including the Moon and other celestial bodies.*", (emphasis added).

understanding of the precise scope and relationship between Art. VI and VII of the OST[45]).

In essence, opinions converge on the idea that the above OST provisions indirectly established a duty for States to adopt measures ensuring that space activities are carried out in accordance with international (space) law. In other words, State parties were made responsible for enacting administrative arrangements concerning the authorization of space activities, as well as for their supervision and smooth completion, as internationally liable internationally liable for any damage that could be caused by space objects under their jurisdiction. Therefore, these articles did not set any substantive requirements for the authorization or supervision of space activities but they still provided the legal basis for the enactment of national legislation [2: 9–10].

In this context, considering the principles and guidelines established in international space law, Art. 3 to 10 of the Greek space law introduced several clarifications: *first,* on the granting of authorization to carry out space activities under Greek law; *second,* on the supervision of space activities by national authorities; and finally, on the conditions for the amendment, suspension or withdrawal of such authorization. It stands to reason that all these rules must be applied in a complementary manner to the provisions of international space law.

5.2.1 Authorization and Control of Space Activities

Many countries have already adopted a specific licensing regime applying to space activities, which allows them to exercise tight preventive control as well as continuous supervision over said activities (especially when conducted by private entities). As a direct result, all "space activities, or activities that fall within the scope of their national space legislation, are prohibited to be conducted unless a license is granted" [2: 146]. However, from a practical perspective, the granting of authorization to conduct a space activity is perhaps one of the most challenging issues to address, given that administrations are required to decide while considering diverging interests; such as the need to encourage and support their domestic space sectors but to also guarantee the safety and viability all actors operating in space.

From a legal viewpoint, there are several types of authorization regimes, as different States adopt different approaches to the procedure [2]. Indeed, a *standard practice* mainly consists of granting an authorization to conduct space activity through a single procedure and under a single set of conditions. A slightly divergent approach to the procedure appears to invovle the establishment of rules providing for *ab initio* exemptions from the standard regime, based on diverse but well-defined criteria, like the importance of the activity or its low degree of hazardousness. For

[45] This important topic remains beyond the scope of the present study (see for more details, *inter alia* [24]).

instance, the UK provides for such exceptions in cases where "a license is not necessary to ensure compliance with the international obligations of the UK" [2: 24],[46] while the Australian Space Activities Act provides that exemption certificates can be granted "for the space license, the launch permit and the permit for return of an Australian-launched space object (*in case*) the conduct is (*inter alia*) required to deal with an emergency" [2: 41].[47]

A *second approach* consists in the establishment of distinct license regimes to authorize space activities of different natures; in this case, "the scope of these licenses differs per license" [2: 33]. Such a multiple license regime has been adopted *inter alia* by Australia (which provides for the granting of a 'Launch Permit', to authorize activities such as the launch of a particular space object or a series of space objects; an 'Overseas Launch Certificate', which is quite similar to the launch permit but applies to space objects launched from a launch facility outside of Australia; and an 'Authorization for the Return of Space Objects Launched Outside Australia' [2: 37–41]); Indonesia (which—to date—differentiates between launching activities and commercial space activities [2: 97–98]) and New Zealand [2: 114–120].

Against this background, the Greek space law opted for a standard procedure, based on the granting of authorization to conduct space activity through a unique process and under a single set of conditions, coupled with a few rules allowing the competent space authority to exercise some discretion where justified. Hence, law n° 4508/2017 established the general principle that *all kinds* of space activities require official authorization from the competent space authority, namely without providing *ab initio* any exceptions for specific persons or activities, or differentiating between space activities according to their purpose.

5.2.1.1 Basic Features of the Authorization Regime Established in Law n° 4508/2017

To begin with, Art. 3 of law n° 4508/2017 stipulates that an authorization—from the competent space authority—is required, to carry out space activity under the jurisdiction of the Greek State. More precisely, Art. 3 reiterates the obligation established in Art. VI of the OST,[48] and clarifies that prior authorization is necessary for the conduct of space activities, *such as* those related to the launching, operating,

[46] Or in case "a person is acting as an employee or agent of another person or for activities in respect of which it is certified by Order in Council that arrangements have been made between the UK and another State to secure compliance with the international obligations of the UK" [2: 24].

[47] "The regulations stipulate that what should be considered when granting an exemption certificate is whether the conduct is required to deal with an emergency; the probability of the conduct causing substantial harm to the public health, public safety or property; and the probability of the Commonwealth being exposed to international liability for damage caused by the conduct" [2: 41].

[48] OST, Art. VI (2nd sentence): "The activities of non-governmental entities in outer space, including the Moon and other celestial bodies, *shall require authorization and continuing supervision* by the appropriate State Party to the Treaty", (emphasis added).

maneuvering and returning of space objects; interestingly though, and contrary to the Australian law, the concept of *return* is not defined in the Greek space law.[49] Be that as it may, using the expression *such as*—in Art. 3 of law n° 4508/2017—seems to indicate that the legislator intended to adopt a broad and future-proof approach, potentially applicable to a wide array of upcoming space activities.

Furthermore, Art. 3.2 (as well as other provisions, such as Art. 5 and Art. 7.1 which reference the authorization procedure) also lays down that it is the Minister of Digital Governance—literally, mention is made in the law to the former competent minister, i.e., the Minister of Digital Policy, Telecommunications and Information[50]—who is the competent authority to grant authorization, at the request of any space operator(s) and in the form of an individual permit. An acute observation mentioned in the Swedish Draft on Space Activities could also be considered relevant to the above rule adopted in law n° 4508/2017, and possibly explain the normative choice made by the Greek legislator: "The decision to keep the power to grant or refuse a license on the governmental level is based on the fact that such licenses might be important from a foreign policy point of view. Furthermore, the number of applications was estimated to be relatively small in the foreseeable future, which makes it unnecessary to delegate the licensing" [2: 18].

As is the case in many countries, it is the space operator—not the manufacturer, as commented upon during the prior public consultation of the draft space law[51]—who is entitled to submit the request. Therefore, a thorough evaluation of the qualifications of each applicant will take place, which nevertheless does not appear to include an in-depth scrutiny of the applicant's repute or any review of their moral or ethical guarantees; contrary to what is established in other national space legislations, like in French law [2: 79].[52]

Following on from that, the authorization to conduct space activity shall be granted *ad personam*, namely to the space operator personally, and clearly refer to a specific space activity programmed to be conducted within a determined or undetermined time period [1: Art. 3 and Art. 4.1]. However, as the authorization is granted personally while a space object's ownership or operatorship may change

[49] The Australian Space Activities Act "defines 'return' as the return, or attempted return, of a space object from beyond a distance of 100 km above sea level to Earth" [2: 34].

[50] Pursuant to Art. 1 para 2 of Presidential Decree n° 81/2019, [25] the Ministry of Digital Governance, which was created at that time, absorbed the main competences of the (former) Minister of Digital Policy, Telecommunications and Information. The power to grant authorization to conduct space activity, established in Art. 3 of law n° 4508/2017, was therefore transferred to the new ministry.

[51] It was underlined, during the public consultation on the draft of law n° 4508/2017, that if an authorization obligation were imposed on manufacturers, the license fee could act as a deterrent for all the research and non-profit institutions that are working on the development and improvement of space infrastructure [13: Art. 3].

[52] "The applicant for a license will need to provide detailed information about the applicant; documents which allow the authority to assess, inter alia, the existence of the moral (. . .) guarantees of the applicant (conditions of good repute and the absence of personal bankruptcy (. . .), quality management systems in the company, qualifications and training policies for staff)" [2: 79].

hands several times during its operation (which is a current practice in the space sector, today), the Greek space legislator additionally took into account (and regulated) cases involving a transfer of space activity [1: Art. 10]. In this instance, the purpose of the Greek legislator was to ensure the sound administrative management of space activities, by considering the specific circumstances under which space operations are carried out.

Finally, Art. 3.3 of the Greek space legislation reaffirmed the fundamental principle enshrined in the OST, and established that all space activities must be conducted in accordance with the principles of international law, as well as with the rules adopted in the international treaties, conventions and agreements [1: Art. 3.3.] i.e., to which Greece is a State party.

Overall, in transposing the obligation of Art. VI of the OST[53] and establishing the practical means of its implementation, the national legislator apparently followed draft Art. 3 of the Sofia Guidelines. Indeed, said provision mentions that "*All* space activities are *subject to authorization*. Authorization shall be granted by the minister of [reference must be made to the competent minister or authority, in each State][54]". At the same time, it was outlined in the comment on draft Art. 3 that "Authorization *may take several forms* ranging from general licenses to individual licenses or permits[55]", which means that States also enjoyed some discretion as regards their choice of measures, [7: Art. 3].

Therefore, to achieve the objective of ensuring effective supervision of the activities taking place within Greek jurisdiction, the procedure to be followed and the specific terms and conditions for authorization—where authorization of a space activity is requested—were laid down in more detail in the subsequent articles, namely in Art. 3 to 7 of law n° 4508/2017.

5.2.1.2 Analytical Presentation of the Substantive Requirements for Authorization

In establishing a national regulatory and supervisory regime for space activities, the Greek legislator took into consideration a number of specific elements. The explicit goal was to efficiently ensure the safe conduct of space operations at all levels, as well as to protect national security, safeguard public health and preserve the safety of goods and persons (in addition to fulfilling the obligations arising from international space law, incumbent upon the Greek State).

---

[53] As underlined, "States have the obligation to authorize and continually supervise space activities, conducted by their nationals in accordance with Article VI of the Outer Space Treaty. (...) states are left free in their choice how to adhere to the obligation to authorize and continually supervise space activities carried on by private entities" [2: 146].

[54] Emphasis added.

[55] Emphasis added.

Practically, the four factors that play a key role in this context are: the characteristics of the space operator (e.g., financial credibility, competence, experience); the protection of the space environment (e.g., pollution prevention); the protection of the national territory and interests (e.g., protection of domestic security and interests, protection of persons) and compliance with the Greek State's international obligations [3: 2, see comments under Art. 3 and Art. 4.].

Within the context of these space policy priorities, more specific requirements—pertaining principally to the space operator and the space activity—were enumerated in detail.

5.2.1.2.1 Requirements Relating to the Applicant (i.e., to the Space Operator)

Under Greek law, the person entitled to submit a request for authorization to carry out a space activity is the space operator. In this regard, Art. 4.1 (α) of law n° 4508/2017 stipulates *first* that the space operator must have *the necessary* credibility, competence and experience to conduct space activities.[56] The article does not provide any further clarifications or guidance as to how the term *necessary*, or even the concept of *credibility*, should be interpreted. The task of addressing this difficult interpretative question will, therefore, most likely lie with the designated competent authority and possibly take the form of a mere administrative practice.

In addition to the above, the space operator must comply with the technical requirements—necessary for the proper use of orbital positions and their corresponding radiofrequencies—that the International Telecommunication Union (ITU) requires space operators to meet [1: Art. 4.1.(στ)]. Finally, they must comply with the insurance requirements established in Art. 11 of the law [1: Art. 4.1.(ζ)] and take all the appropriate measures to ensure the smooth and safe conclusion of the space activity [1: Art. 4.1.(η)]. Law n° 4508/2017 does not specify the precise content of this kind of measures; it is, therefore, likely that their effectiveness, efficiency and relevance will be evaluated by the competent authority on a case-by-case basis, taking into consideration the best practices in this field.

The Sofia Guidelines, in draft Art. 4, provided a major reference point for national legislators, including the Greek legislator, with respect to key substantial conditions that must be met for granting a space activity authorization. In particular, they suggest that the competent space authority must ensure that applicants have indeed the financial and operational capacity to efficiently complete the proposed space activity. Literally, the instrument mentions that an authorization should be granted under the conditions that the operator: (i) is in a financial position to undertake space activity [7: draft Art. 4.1.(a)]; (ii) has proven to be reliable and to have all the required technical knowledge [7: draft Art. 4.1.(b)]; (iii) has complied with ITU Regulations on frequency allocations and orbital positions [7: draft Art. 4.1.(h)] and

[56] Emphasis added.

(iv) complies with insurance requirements as is determined in Art. 12 [7: draft Art. 4.1.(i)]. Therefore, a comparison between draft Art. 4 and the relevant article of the Greek space law shows that the latter is, in essence, closely aligned with draft Art. 4 of the Sofia guidelines, as it only adds a very specific rule on the smooth and safe conclusion of the space activity.

5.2.1.2.2 Conditions Applying to the Proposed Activity

In addition to the conditions applying to the space operator, Art. 4.1 (β) to (ε) of the Greek space law clarifies that the proposed space activity must, in itself, meet specific criteria to be authorized. Indeed, it should not constitute—or be related to, or conducive to—a threat to national security, public policy, the security of persons and property, or public health [1: Art. 4.1.(β)]. Furthermore, the activity should not cause, or even be likely to cause, environmental damage to the space and/or Earth environment, as stated in Art. IX of the OST.[57]

Following this reasoning, Art. 4.1. (δ) of the Greek space law clearly stipulates, *first*, that specific provision must be made for space debris and/or residues mitigation and management, in line with the technological state of the art as well as international practices. Be that as it may, the terms *technological state of the art* and *international practices* are not further defined; hence, they most probably refer to widely accepted standards and rules used in the space industry sector. *Second*, Art. 4.1 (ε) reaffirms that the proposed activity must not cause any pollution of outer space or celestial bodies, or create any other negative changes to the space environment, and/or contamination or environmental damage to the Earth. In reality, a large number of States have included specific provisions in their national space laws, to protect the space environment[58] (the particular issue of space debris mitigation is likewise addressed but "often receives separate attention [2: 177]" and is mainly regulated at the international or regional level). Finally, Art. 4.1. (γ) of law n° 4508/2017 stipulates that the proposed space activity should not conflict with any obligations of the Greek State under international law, or with any strategic plans reflecting Greek foreign policy.

[57] OST, Art. IX establishes that "In the exploration and use of outer space, (...) States Parties to the Treaty shall be guided by the principle of cooperation and mutual assistance and shall conduct all their activities in outer space, including the Moon and other celestial bodies, *with due regard to the corresponding interests of all other States Parties to the Treaty*. States Parties to the Treaty shall pursue studies of outer space, (...) and conduct exploration of them so as *to avoid their harmful contamination and also adverse changes in the environment of the Earth* resulting from the introduction of extraterrestrial matter and, where necessary, shall adopt appropriate measures for this purpose. (...)", (emphasis added).

[58] For example, comparative law shows that the UK, Australia, China, Belgium, the Netherlands, France, Austria, Indonesia, Denmark and New Zealand have all included requirements with respect to the protection of the environment as a condition to the space activity license [2: 173–176].

All the requirements applying to space activities mentioned above are again aligned with the ones established in draft Art. 4 of the Sofia Guidelines, laying down that proposed space activities: (i) should not cause environmental damage to the Earth or outer space [7: draft Art. 4.1.(c)]; (ii) should be undertaken in such a manner as to mitigate to the greatest possible extent any potential space debris in accordance with draft article 8 [7: draft Art. 4.1.(d)]; (iii) should be compatible with public safety standards [7: draft Art. 4.1.(e)] and (iv) should not be against national security interests [7: draft Art. 4.1.(f)], international obligations and foreign policy interests of the State [7: draft Art. 4.1.(g)]. Importantly, the Greek legislator decided to substantially broaden the scope of application of the corresponding article on authorization requirements, by explicitly including—in the Greek space law—the protection of the health and safety of persons; interestingly though, a similar provision may be found in the UK Space Industry Act, which introduced "the requirement that the applicant for a license must carry out an assessment of the risks to the health and safety of individuals who will take part in the activities to be authorized by the license" [2: 26], as well as in Chinese law [2: 50].[59]

Pursuant to Art. 4.2 of law n° 4508/2017, the above general requirements may be supplemented, following a decision of the competent minister, to ensure the security of persons and property, protect the environment, ensure the optimal use of the atmosphere and outer space, and/or preserve the Greek State's strategic, geopolitical, and financial interests. The competent minister may at the same time impose additional or specific requirements for the granting of a particular authorization, on a case-by-case basis.[60] Finally, the conditions and duration of an authorization may be modified, but only by a duly justified decision of said authority [1: Art. 4.4].

As already underlined, the Greek legislator clarified—in Art. 4 of law n° 4508/2017—that account must be taken of the current technological and scientific standards (i.e., the law refers to the *technological state of the art*) when assessing the overall viability and feasibility of the proposed space activity; which brings a certain degree of flexibility and enables the competent authority to adapt the rule to each specific case. As a corollary, the deciding authority may, at their discretion, request technical assistance from third parties, such as from Greek or foreign (space) experts, qualified advisers and (space) scientists [1: Art. 4.3]. A comparative legal analysis shows that, unsurprisingly, comparable provisions have been adopted in other legal systems[61] as well.

[59] In order to obtain a license under Chinese law, the applicant is required to establish that the proposed space activity will not cause irremediable danger to public health, safety or property, due to major negligence or intentional acts [2: 50].

[60] Within this purpose, the minister may set conditions applying e.g. to the space activity location, the operator's main establishment location, or a certification of compliance with the insurance requirement applying to the space activity and space object [1: Art. 4.3].

[61] For instance, in Belgian law "The Minister can further supplement the information by requesting substantiated advice on the basis of juridical, technical and economic criteria" [2: 60].

5.2.1.2.2.1 Formal Requirements - Information Disclosure: The Application Dossier

Although the content of the application dossier for a space activity authorization—including any relevant information which must be provided to the competent authority for supervision purposes—is of fundamental importance, international law does not specify such details. As a result, the Sofia Guidelines attempted to provide practical clarifications to national legislators as mentioned above, to assist them in implementing States' obligation to set up a supervisory regime for space activities under their jurisdiction.

More precisely, the comment under draft Art. 4 of the Guidelines entitled *Conditions for authorization* reads: "It must be made clear that any applicant is personally reliable and in a secure financial position. Moreover, requirements of foreign policy, national security, public safety, international telecommunication regulations and insurance should be fulfilled. *Written documentation* gives proof of these facts[62]". In other words, the national legislators were on the one hand encouraged to adopt a procedure based on written supporting documents but, on the other hand, left free to decide on the specific content of the submission dossier, as long as the submitted documents provide proof of the reliability of the space operator and compliance with all the requirements pertaining to the activity.

Against this background, Art. 5 of the Greek space law drew up a list of documents which must be submitted to the competent national authority, for the authorization of a space activity. Nonetheless, to have a comprehensive view of the rules applying in this case, Art. 5 should be read in conjunction with Art. 6 which clarifies the types and content of the environmental impact assessment (EIA) documentation which should also be included in the application dossier.

5.2.1.2.2.1.1 Content of the Application Dossier: Art. 5 of Law n° 4508/2017

As mentioned above, any request for authorization to conduct a space activity under the jurisdiction of the Greek State must be submitted by the space operator(s). In other words, it is the space operator (i.e., who could be either a natural or a legal person, as explicitly mentioned in Art. 2.2 of law n° 4508/2017) who is fully responsible for collecting, preparing and finally addressing the necessary documentation to the competent authority. Hence, following on from that, Art. 5 enumerates all the information that must be provided when submitting the application, although it is apparent that this list may be complemented, in case that the conditions for granting an authorization are strengthened by a decision of the competent minister, as provided for in the law and presented above [1: Art. 4]).

More precisely, Art. 5.1 establishes that a demand for authorization to carry out a space activity shall be submitted by the space operator, and that it must be *co-signed by any other participants* in the activity concerned, as well as by any co-owner of the

[62] Emphasis added.

space object. As the law does not provide clarifications on who qualifies as a participant in the space activity for the purposes of this article, this issue will most probably be regulated taking into account practices in the field. In this context, Art.5 explained *first* the obligation to pay an application fee and the necessity to prove the feasibility of the space activity, as explained below.

– ***Payment of an application fee***

Art. 5.1 of the Greek space law mentions that the applicant has the obligation to pay a fee amounting to six thousand (6.000) euros for submitting a dossier to the competent space authority, namely to the Minister for Digital Governance. Interestingly, in its initial version, the article was rather vague, as it included a general reference to the possibility that the fee amount could also be adjusted following a decision by the competent Ministers.[63] Indeed, the draft version of this provision did not include any additional details on the conditions under which such a decision could be made. In this context, it appeared that the competent authority would have too wide a margin of discretion, without being bound in any way to effectively adjust the fee for any specific purpose.

The reaction was evidently immediate, and it was underlined—during the consultation process—that any exemptions benefiting space activities, such as non-lucrative ones, should be clearly defined. The payment of a six thousand (6.000) euros fee was regarded as being, in such cases, disproportionate, burdensome or impractical, while it was argued that it would act as a disincentive for non-lucrative space activities. In addition to that, it was noted that when determining any fee amount, account must be taken of the fact that several private companies involved in space-related activities are often small and medium size businesses which may not be able to pay a fee of six thousand euros per application [13: Art. 1]. Hence, the suggestion was made to allow for the adaptation of this amount, considering the specific purpose of a space activity as well as the financial situation of the applicant [13: Art. 3 and Art. 5].

The comment was taken into consideration, and an amendment was introduced in the relevant part of the draft. As a result, Art. 5 of law n° 4508/2017 contained a clear exemption—on this specific point—to the benefit of non-lucrative space activities and activities of public interest, bearing in mind the proliferation of low budget and/or research activities *inter alia*. The rationale was that such activities are indeed burgeoning today, thanks to the growing use of small satellites, which are particularly suitable for space operations requiring speed, simplicity and productivity for better competitiveness. In this sense, their low cost coupled with their flexibility of operation has made them increasingly popular within the space scientific and

[63] The draft law n° 4508/2017, Art. 5.1, initially read: "The application for a space activity authorization shall be submitted by the operator to the Minister of Digital Policy, Telecommunications and Information, accompanied by a fee of six thousand (6.000) euros, *the amount of which can be adjusted by decision of* the Ministers of Finance and of Digital Policy, Telecommunications and Information", (emphasis added) [13].

research community as well,[64] particularly in emerging space-faring countries with more limited resources to invest in space activity.

Accordingly, the Greek space legislation initially provided that the competent minister (today, Minister of Digital Governance) exempts applicants from the obligation to pay the fee provided for in Art. 5.1, in case a space activity is intended—exclusively or in part—for public purposes or exclusively for non-commercial research, experimental and educational purposes [1: Art. 5.1]. However, the rule was recently amended. Specifically, Art. 39 of law n° 5099/2024 reintroduced the discretion of the deciding authority, even though, from a comparative law perspective, it is not unusual for research space activities to benefit from a specific derogatory regime.[65]

– ***Control of the operator and examination of the project's feasibility***

In addition to the above, the national legislator explictly laid down the requirement that the competent space authority must examine the feasibility of the proposed space activity, in detail and from all angles. Hence, both the reliability of the space operator(s) and the feasibility of the activity must be proven through a set of supporting documents, which ought to be attached to the application. All these documents are enumerated in Art. 5.2 of law n° 4508/2017, and aim to identify the operator(s) and the legal form of the person or entity concerned [1: Art. 5.2.(α)]; provide information on the operator's activities and situation, in addition to their capacity to fully guarantee the financial viability of the proposed space activity submitted for authorization [1: Art. 5.2.(β)]; as well as, finally, give a complete and accurate description of the activity and the space object for which the authorization is being sought [1: Art. 5.2.(γ)].

Moreover, the applicant must submit full identification documentation as well as all relevant clarifications on the manufacturer of the space object [1: Art. 5.2.(δ)] and on the natural and/or legal persons on behalf of whom the space activity will be carried out [1: Art. 5.2.(στ)]; full contact details of the persons designated by the operator as their contact points [1: Art. 5.2.(ζ)]; the insurance contract [1: Art. 5.2.(η)][66] and, finally, the information required for registration in the National Register for space objects [1: Art. 5.2.(θ)]. In addition to that, it is clearly laid down that the

[64] "(...) the Smallsat market is booming, which can be confirmed not only by the number of Smallsats launched, but also by the increasing amount of conferences in this area. What began as scientific and research activity is now also catching interest of private initiatives (...). Furthermore, there is a perspective that there will be a boom in the Smallsat market whose realization will depend on several factors" [26: 270–271 and 273].

[65] Such as, for instance, in Austrian law: "The insurance regime set out in the Act can be differentiated in three categories (...) The second category can be exempted from having insurance or have the required insurance limited by administrative decisions. However, this applies only if the space activity is conducted in the public interest, meaning the activity serves science, research or education." [2: 91].

[66] This requirement is further developed in law n° 4508/2017, Art. 11.

applicant should also submit any other information and supporting document that they consider necessary and relevant to support their request [1: Art. 5.2].[67]

In general, Art. 5 of the Greek space law neither prohibits nor explicitly mentions very specific types or categories of (official) documents, but rather focuses on the facts that need to be proven (which does not exclude referencing particular types of documents where necessary, such as in Art. 5.2.(ε)[68] or Art. 5.2.(η)[69] of law n° 4508/2017). In this sense, the provision remains essentially flexible and adaptable to the particularities of each space activity; especially considering that this kind of activities are often based on projects that—frequently—involve entities and persons (such as groups, consortia etc.) of different nationalities and legal systems. These participants will most probably only be able to provide the kind of official document (s) which may be drawn up in their own country, to support their claim. From a certain perspective, this view is supported by the general reference—in Art. 5.2, last sub-paragraph—to the opportunity given to the applicant(s) to also submit "any other information or supporting document *he considers relevant* in the justification of his request[70]", a rule that was undoubtedly adopted to assist both the competent authorities and the parties concerned in finding acceptable solutions.

5.2.1.2.2.1.2 Measures Aimed at Environmental Protection in the Greek Space Law

It has been argued that "(w)hen it comes to the protection of the environment, there are no provisions under international space law that stipulate defined obligations with respect to the environment. However, Article IX of the Outer Space Treaty does state that: State Parties to the Treaty shall pursue studies of outer space, including the Moon and other celestial bodies, and conduct exploration of them so as to avoid their harmful contamination and also adverse changes in the environment of the Earth (…) and, where necessary, shall adopt appropriate measures for this purpose. Furthermore, through Article III of the Outer Space Treaty, states are also bound to international environmental law when authorizing space activities by their nationals" [2: 173].

Put differently, scholars contend that, *on the one hand*, international space law does not lay down a clear and binding framework for environmental protection in space, and thus no precise rules can be derived from it (or from the OST in particular) on space sustainability as such [27]. In this context, mention is made, *inter alia*, of the "narrow applicability" of Art. IX of the OST (due to its "express confinement to

[67] Law n° 4508/2017, Art. 5.2 - last sub-paragraph.

[68] According to Art 5.2.(ε) of Law n° 4508/2017, the applicant (i.e., the space operator) must precisely submit a declaration of honor [*Υπεύθυνη Δήλωση*] that will attest to their compliance with all the environmental requirements and practices.

[69] Following Art 5.2.(η) of Law n° 4508/2017, the applicant (space operator) must also provide a copy of the insurance policy.

[70] Emphasis added.

the 'studies' and 'exploration' of space"), since "the clause focuses primarily on the release of harmful microbiological organisms of terrestrial origin from any accidental collision or explosion in outer space, and it is questionable whether or not the definition extends to include all other sources of environmental harm" [28].

On the other hand, it is also argued that Art. IX of the OST may be regarded as a first milestone in the path towards the protection of the outer space environment *lato sensu*, as it did create a general though unequivocal obligation on States. More precisely, despite its vague and broad wording, Art. IX of the OST is said to have at least established "a 'proscriptive positive legal obligation' for States to (i) avoid harmful contamination of celestial bodies and (ii) undertake international consultations in advance before any potentially harmful interference may arise from their activities" [28]. Hence, following this line of reasoning, a clear duty was created on the basis of Art. IX, even though the responsibility for laying down more precise rules on this specific matter belongs to the States.

In more practical terms, it is suggested that, given the growing awareness of (space) environmental protection—coupled with the emergence of an Environmental Impact Assessment duty [29] as a part of international environmental law—, national legislators should ensure that space activities are designed and implemented following the conduction of an appropriate Environmental Impact Assessment (EIA) or other similar measures (as is the norm for many activities conducted on Earth) [28]; especially in case there is a risk that they may have adverse impacts on the environment *lato sensu*. The ultimate goal is to evaluate the environmental impact of the proposed space activities before their implementation.

As a matter of fact, draft Art. 4 of the Sofia Guidelines mentions that a space activity should be authorized on the explicit condition that "The space activity *does not cause environmental damage to the Earth and outer space* in accordance with article 7[71]" [7: Art. 4.1.(c)]. Be that as it may, draft Art. 7—i.e., which is referred to in draft Art. 4—, only lays down the minimum measures needed to achieve the objective of environmental protection. In particular, the provision *first* reaffirms that the space activity must not cause environmental damage (to the Earth, outer space, or parts thereof), either directly or indirectly [7: Art. 7.1]. Essentially, however, it additionally stipulates that "(a)n environmental impact assessment *is required before the beginning* of a space activity" [7: Art. 7.2][72], leaving the precise regulation of these matters to national legislators.[73]

Quite importantly, comparative studies on national space laws show that the majority of States have, to date, adopted measures aimed at ensuring the protection

[71] Emphasis added.

[72] Emphasis added.

[73] Sofia Guidelines, draft Art. 7.3: "Details of the environmental impact assessment shall be laid down in an implementing decree/regulation".

of the space environment, either by specifically requiring an EIA, [30][74] or by stating a general but clear obligation for space operators not to adversely impact the environment while carrying out space activities.[75]

Hence, in reference to the choices made by national legislators on this matter, it appears that "a general trend can be observed with respect to the provisions regulating environmental protection, namely, that the provisions are very broad and usually only stipulate the obligation to conduct an assessment and prevent harmful contamination. Such provisions can be split into two categories. The first category contains provisions that are concerned with the protection of the Earth's environment and the outer space environment. The second category consists of provisions that specifically focus on the mitigation of space debris" [2: 173–176]. The Greek space law established both kinds of rules.

– ***The obligation to submit an EIA following Art. 6 of law n° 4508/2017***

According to the Greek space law, the applicant must submit [1: Art. 5.2.(ε)][76] a declaration of honor [*Υπεύθυνη Δήλωση*] attesting to their compliance with all the environmental requirements and common practices, as well as with all the national and international environmental legislation in force,[77] as this obligation is further described in the law. Indeed, more substantively, this declaration must be accompanied by an Environmental Impact Assessment (EIA) containing measures for environmental protection, such as contingency and mitigation measures. This obligation is developed in Art. 6, which explicitly underlines that for every single space activity, the operator must submit—at regular intervals—a report on any environmental impacts potentially caused by the space object, to the competent minister (i.e., today the Minister of Digital Governance).

Interestingly, regarding *first* the standards that must be used to prepare the EIA, law n° 4508/2017 remains unclear. However, the Explanatory Memorandum of the Greek space law highlights that the specific form and content of the EIA to be submitted by applicants to the competent authority may be clarified in the future. At the same time, considering that the purpose was to provide—right from the outset—a comprehensive and reliable assessment of said risks to the competent space

[74] "The legal frameworks of Belgium and France are exceptions as they require that EIA considers extraterrestrial impacts. Belgium's Law on the Activities of Launching, Flight Operation or Guidance of Space Objects (Kingdom of Belgium 2013) requires that an EIA be submitted prior to the launch, assessing the effects of the action on both the Earth and any celestial body affected [31] etc." [30].

[75] "The UK stipulates that a condition can be set to prevent the contamination of outer space or adverse changes in the Earth's environment. Likewise, China, the Netherlands, Austria, Indonesia, Denmark and New Zealand have stipulated the obligation that activities need to be conducted in an environmentally safe manner, avoid pollution and should not cause harmful contamination or adverse changes in the environment." [2: 176].

[76] In addition to the other documents and information mentioned in Art. 5.2, namely Art 5.2.(α) to 5.2.(δ), and 5.2.(στ) to 5.2.(θ) of the Greek space law.

[77] Based on this Declaration of Honor, the space operator shall be held liable for the accuracy and completeness of the reports [1: Art. 6.7].

authority, it is mentioned that until such clarifications on the official procedure for preparing the EIA are released, the international and European standards applying to this specific matter, as well as those used in the elaboration and evaluation of such reports in European practice, may be relied upon [3].

Therefore, for the sake of clarity and coherence, the Explanatory Memorandum explicitly referenced tools that could and should be taken into account (such as the International Standardization Organization (ISO) 24113; the European Cooperation for Space Standardization (ECSS) Space Sustainability; the European Space Agency (ESA) Space Debris Mitigation Compliance Verification Guidelines etc.), while also suggesting the consideration of relevant standards adopted in satellite manufacturing countries such as the US, France, Russia etc. Finally, the Explanatory Memorandum clarified that in some instances, although a scientific and/or research space activity may not meet all the requirements and standards of the applicable regulations, it could still be authorized [3: 3, see comment under Art. 6.]. Remarkably, none of these clarifications or legal instruments were finally mentioned in law n° 4508/2017, which appears to give the competent authority a certain—however, in practice quite limited—flexibility in assessing the EIA, taking into account national circumstances as well as current practices in the space industry.

On the other hand (*second*), the Greek space law established quite specific rules on the frequency and content of reporting for each stage of the space operation—i.e., reports to be submitted by the space operator to the competent authority—, to enable the authority to carry out its control and monitoring duties. More precisely, Art. 6 of law n° 4508/2017 laid down the obligation for the space operator to communicate three types of EIA to the competent authority, on a regular and systematic basis.

By way of explanation, a *first report* (that is, an EIA) must be prepared prior to the granting of authorization to conduct the activity, assessing any potential effects on the Earth or the outer space environment that might result from (or be related to) the launching or operating conditions of the space object [1: Art. 6.2]. After the launching of the object—or during its operation—, a *second (or interim) report* must be submitted, containing an assessment of the effects which have occurred and significantly impacted the Earth or the outer space environment [1: Art. 6.3]. Lastly, a *third and final report* must be presented following the return of the space object, or alternatively after the completion or the cessation of the space activity [1: Art. 6.4].

Finally, Art. 6.1 provides for the possibility to appoint one or more (space) experts to evaluate and review all these EIA reports, to the benefit of the Minister of Digital Governance and any other Minister(s) that would be co-competent for the assessment of an application dossier.[78] It is noteworthy that Art. 6.6 precisely mentions that space operators shall bear the costs related to the preparation of all these reports, even though imposing on applicants the burden of covering the particularly high expenses incurred through such official controls—as a general

[78] Importantly, the law mentions that compliance with the relevant standards is sufficient in order to prove observance of the environmental assessment (and authorization) requirements [1: Art. 6.5].

rule—raised significant concerns within the space industry, on the grounds that this practice is inherently unfair and damaging to the space industry's interests [13: Art. 14].

– ***The general obligation to provide for space debris mitigation***

It is remarkable that, regarding environmental protection, a few States have additionally established a specific obligation to submit a space debris mitigation plan. In particular, such is the case with New Zealand's space law[79] or France's space law, which clearly provides that "(t)he EIA needs to consider the operation of the rocket engines (...) as well as the production of space debris and the transport of radioactive material on the launch vehicle. The results of this EIA need to lead to appropriate measures to limit the identified risks, which inter alia, include a plan on the prevention of environmental damage and *a plan on limiting space debris*. As far as the mitigation of space debris is concerned, it is further elaborated that the launch system must adhere to a number of technical requirements" [2: 85].

Literally, the Greek space law does not regulate the matter of debris mitigation in detail. Be that as it may, the obligation to provide for space debris management is unequivocally established in para. (δ) of Art. 4.1, laying down the general conditions for authorizing a space activity. In fact, said article explicitly mentions that the operator must make provisions for debris mitigation measures (in addition to fulfilling all the other conditions for authorization), and that these measures must be prepared in line with current practices in this field. At the same time, Art 4.1.(ε) strengthens further this obligation by reaffirming that the space activity must not cause any pollution in space or on celestial bodies, or other hazards or adverse effects [1: Art. 4.1]. Finally, the obligation to provide for space debris mitigation as a clear condition of approval, is mentioned in the Explanatory Memorandum [3: 2, see comment under Art. 4].

Overall, potential weaknesses of this provision are that, *first*, the key concept of space debris is not clearly defined; this is, however, not truly surprising given the fact that "there is as yet no universally accepted or legally binding definition of orbital space debris".[80] In addition to that (*second*), no further specific rules were established in the Greek space law on the preparation of the space debris

[79]"(...) the licensing does consider environmental impact and space debris. All licenses require the minimization of contamination of outer space or adverse changes in the Earth's environment. In addition, for those licenses and permits that concern actual activities in outer space, a space debris mitigation plan is required. The requirements for such a mitigation plan are stipulated in the Licenses and Permits Regulations" [2: 123].

[80]"While there is as yet no universally accepted or legally binding definition of orbital space debris, COPUOS has developed a definition, incorporating debris both in Earth orbit but also in the process of "de-orbiting," as follows: "[A]ll man-made objects, including fragments and elements thereof, in Earth orbit or reentering the atmosphere, that are non-functional." This definition itself comes from various non-binding instruments developed through COPUOS dealing specifically with space debris mitigation" [32: 378].

management measures provided for in Art. 4.1.(δ), which could lead to legal uncertainty regarding the precise measures that should be adopted for debris mitigation.

In fact, regarding this latter point, it is noteworthy that the Sofia Guidelines elaborate extensively on the matter of space debris mitigation, contrary to the Greek space law. In particular, the Guidelines first suggest the adoption (on the part of the national legislator) of the fundamental obligation on the space operator to avoid the creation of any space debris, as stated in draft Art. 4.1(d)[81]; this provision largely corresponds to the content of Art. 4.1.(δ) of law n° 4508/2017. Additionally, it is proposed (in draft Art. 8[82]) that a set of specific measures could be adopted to address the space debris challenge. In particular, it is clarified that the obligation established in draft Art. 8.1 (mentioning the duty to mitigate to the greatest possible extent any potential space debris when carrying out space activities) should include the obligation "to limit debris released during normal operations, to minimize the potential for in-orbit break-ups, to prepare for post-mission disposal and to avoid in-orbit collisions in accordance with international space debris mitigation standards" [7: Art. 8.2].

Hence, law n° 4508/2017 does not transpose *verbatim* the detailed wording of the Sofia Guidelines, in draft Art. 8. However, Art. 6.5 of the Greek space law establishes clearly that a joint ministerial decision[83] shall determine the specific content of all the reports, as well as the requirements and technical standards that the operator's activity must comply with. At the same time, pending the adoption of such a decision, the existing national, European and international standards, along with all relevant good practices, shall apply. On this basis, although it is not as detailed as draft Art. 8 of the Sofia Guidelines, it appears that Art. 6.5 of the Greek space law ultimately provides for the application of all commonly agreed best practices and instruments relevant to debris mitigation.

Finally, as mentioned above, Art. 6.7 of the Greek space law dictates that the space operator must sign a declaration of honor attesting to their compliance with international practices and environmental requirements in general. This clause seems to establish an obligation of results (i.e., to effectively comply with said rules), rather than a lesser obligation of means.[84] In other words, it appears that this provision obligates the competent authority to examine the effective compliance with all good

---

[81] Sofia Guidelines, draft Art. 4 entitled *Conditions for authorization*, para. 1(d), reads: "The space activity is undertaken in such a manner as to mitigate to the greatest possible extent any potential space debris in accordance with article 8".

[82] Sofia Guidelines, draft Art. 8 is precisely entitled *Mitigation of space debris*, and dedicated to this matter.

[83] Namely, a Joint Ministerial Decision of the Minister of Digital Policy, Telecommunications and Information (now, Minister of Digital Governance), and the Minister of Environment and Energy [1: Art. 6.5].

[84] The obligation of results is directed at guaranteeing the attainment of a particular result, while the obligation of means principally consists of the employment of a duty of care in performing an obligation [see *inter alia*, 33].

practices on space debris mitigation—i.e., by the space operator—, who must duly consider protocols for collision avoidance, practical standards of safety, and guidelines referring to the use of space-based technologies.

5.2.1.2.2.2 Application Assessment and Authorization Approval

Following submission of all the relevant documents and information, the competent authority shall examine the application and make its licensing decision. The procedure —a thorough and detailed assessment of the request for authorization submitted by the space operator—is established in Art. 7 of law n° 4508/2017. Therefore, pursuant to Art. 7.1 as currently in force, the Minister of Digital governance is the competent authority to decide on the approval or rejection of a space activity authorization request, provided, evidently, that all the steps required to that end, and mentioned in Art. 3 to 6 of law n° 4508/2017, have been completed.

In particular, as Art. 7 of the Greek space law mentions, the Minister receiving the application is expected to decide on the request within a period of three (3) months from its submission. Adopting this time-frame was arguably regarded as well-adapted to the needs of the space industry, while also giving the competent authority sufficient time to decide whether the space activity should be approved, rejected or amended [1: Art. 7.1].

In principle, all the information and documents that must accompany an application for authorization are those enumerated in Art. 5.2. Nonetheless, as the authorization is granted on a case-by-case basis, the competent minister is also entitled (pursuant to Art. 7.2) to request additional information[85] or to ask for a thorough investigation of the whole dossier, by recognized experts (as explained below).[86] In this case, the time limit for deciding on the approval or rejection of the application for authorization may be extended for a period of up to three (3) months [1: Art. 7.2].

At the same time, as the authorization is granted for a particular space activity—i.e. the one that must be accurately described in the application dossier—, law n° 4508/2017 lays down that the Minister of Digital governance shall issue an authorization, that may as well contain any additional specific terms and/or conditions that they consider appropriate, [1: Art. 7.1] as justified by the specificity of the activity referred to in the application.

In examining the dossier, particular attention will be given to the measures taken by the space operator to comply with the standards in force, all of which must be accurately described. Based on this content, the competent authority may ask the applicant for clarification on any documents or information provided [1: Art. 7.2].

[85] As provided for in Law n° 4508/2017, Art. 5.4.

[86] Reference is arguably made to experts in the space sector. Interestingly, a clause was included in the Draft of the Greek space law to also utilize defense systems experts; this provision, initially included in the draft version of Art. 5.5, was omitted in the final text of law n° 4508/2017 [13: Art. 5.5].

Failure to provide clarifications within the assigned time period can, in fact, be regarded as sufficient grounds to reject the application for authorization [1: Art. 5.4, in conjunction with Art. 7.3]. The Explanatory Memorandum to the draft of law n° 4508/2017 indicates that this provision was adopted following the reasoning that failure to provide adequate explanations could prove detrimental to national interests, such as security, defense, international relations and/or the protection of the Greek State's rights in outer space [3: 2, comment under Art. 5].

Furthermore, as noted above, considering the importance and complexity of space topics, the competent minister may request (on the basis of Art. 5.5) assistance from experts—natural or legal persons—with significant international expertise on space matters, to receive advice on the operator's reliability and experience; on whether their financial position allows them to successfully carry out the proposed activity; on the manufacturer's reliability and on the ability of both to comply with international good practices and rules. Once appointed, all experts are granted access to the premises, buildings and materials that will be used by the space operator within the framework of the proposed space activity. In return, the experts acting on behalf of the competent authority are bound by an obligation of professional secrecy, and are obliged to ensure the confidentiality of any data gathered during their investigations [1: Art. 5.5]. A refusal by the space operator to grant access to any of their installations *lato sensu*[87] will be regarded as suspect and considered sufficient grounds for rejecting the application for authorization [1: Art. 5.5].

Finally, as laid down in Art. 14 of law n° 4508/2017, the applicant is responsible for paying all the costs associated with the use of space expert services by the competent minister, such as analyses, expert opinions, consultancy, and/or other related costs. In case the operator fails to cover these costs, the minister may—at their discretion—reject the application. Importantly, if an operator who communicates their refusal is already the holder of a license (since the Greek space law, adopted in 2017, also applied to space activities that were already ongoing), the Minister of Digital Governance may decide to temporarily suspend their authorization.[88]

Given the strong interplay between the different areas of the economy—such as the space industry and research activities, as well as between these sectors and environmental protection—, the application dossier will be also communicated to the other minister(s) mentioned in the law.[89] In particular, Art. 5.6 lays down that (in this case) any of the other interested minister(s) informed about the application for a space activity authorization, has/have the right to raise objections within

[87] Law n° 4508/2017, Art. 5.5, only makes a general reference to a "denial of access" which, however, must most probably be read in conjunction with the terms used in the previous sentence of the same article, namely "premises, buildings and materials".

[88] Still, this decision is subject to the discretion of the competent authority, as clearly established in the Explanatory Memorandum to the Draft law [3: 4, comment under Art. 14].

[89] Law n° 4508/2017, Art. 5.6, makes reference to the former Ministry of Finance and Development; to the Ministry of National Defense; to the Ministry of Education, Research and Religious Affairs; to the Ministry of Foreign Affairs and to the Ministry for the Environment and Energy.

40 days of receiving the application dossier; failing this, their agreement shall be presumed [1: Art. 5.6.].

Finally, according to Art. 7.4, once a request for authorization has been approved by the competent authority, the decision granting such authorization shall be published in the Greek Government Gazette.

5.2.1.2.2.3 "Unfavorable" Decisions for the Applicant: The Concerns Expressed

During the public consultation on the draft of the Greek space law, specific concerns were raised due to the rules proposed in Art. 5.4 and 5.5 (establishing the obligations to provide additional information and/or to grant access to the operator's installations, to the competent authority, following its request), as well as in Art. 7.3 of law n° 4508/2017 clarifying that in case no information is received after the expiration of the additional period established in Art. 7.2, the application will be rejected. Interestingly, such reactions were expressed despite the fact that comparable rules have been adopted in other national space legislations[90] too.

First of all, it was suggested that authorization for a space activity should not only be granted following an explicit and formal decision; authorization should also be given tacitly, on the basis of an implicit decision that would result from the silence of the competent authority (i.e., an implicit approval) [13: Art. 7]. This proposal was intended to mainly facilitate the decision-making process to the benefit of operators, and make the procedure faster and easier for them with the view to attracting more private sector participation.

While this approach is quite understandable, such rule could not have been regarded as acceptable, as the States are responsible—under international space law—for the smooth conduct and supervision of all space activities. Hence, their approval of the conduct of each space activity must be clear and unambiguous, as it is in the State's interest to ensure that each operation meets all the safety and environmental standards. In this context, failure to reach a decision within the stipulated time limit could not be equated with an effective authorization that would *ipso facto* make the Greek State responsible for the conduct and supervision of a space activity that would not have been adequately controlled for in detail.

Following on from that, particular mention was made to the fact that by granting a right of access to the installations and materials used for a space activity—which are, in fact, often subject to high secrecy—to allow the competent authority to conduct inquiries on its own initiative, the operator may be confronted by situations where there is a conflict of interests (e.g., the use of such materials is usually subject to strict rules, such as controls on the exports of high technology). Hence, it was proposed that said rule should be reviewed, to allow space operators meet their own contractual obligations [13: Art. 5 and Art. 8]. To answer all these concerns, a specific

[90] *Inter alia*, the Belgian Act on Space Activities mentions that: "The Minister can further request the applicant to disclose any other information deemed necessary; refusal of such a request is a ground for refusal of the application" [2: 60].

protective clause (in Art. 5.5 of the Greek space law) was introduced, mentioning that the experts hired by the competent authority to control the operator's premises and documentation will in turn be bound to ensure the secrecy and confidentiality of all received information.

Regarding this latter point (i.e., the right of access to the space operator's premises), a comparative law analysis shows that the adoption of an obligation to grant such access to the premises seems to be the dominant approach; the rationale for this perspective is—again—to allow the State give its unequivocal approval, in full knowledge of all the facts and conditions under which the space activity will be carried out. Thus, by way of illustration, this viewpoint was adopted in the UK Space Act;[91] the Australian Space Activities Act;[92] the French Act on Space Operations;[93] the Austrian Outer Space Act[94] and the Danish Outer Space Act,[95] *inter alia*, for evident practical (and legal) reasons.

5.2.2 Conduct and Supervision of Space Activities

Art. VI of the OST explicitly mentions that "States parties to the Treaty shall bear international responsibility for national activities in outer space (. . .) and for assuring that national activities *are carried out in conformity with the provisions set forth in the present Treaty*[96]". Therefore, according to international space law, all States need to assume their responsibilities and take—following the granting of an authorization to conduct space activity—all the necessary measures to adequately monitor and control space operations under their jurisdiction, during all the planning and

[91] "Supervision of the activities conducted under the Act is regulated through the inclusion of a condition in the license that can require the licensee to permit inspection of the licensee's facilities, inspection and testing of the licensee's equipment and permit the inspection and taking of copies of documents relating to the information required to be given" [2: 30].

[92] "The supervision of the activities that fall within the scope of the Act has been regulated (. . .), through the inclusion of a provision that stipulates the obligation to furnish any information that is requested to the Minister. Furthermore, the holder of the space license must allow reasonable access to the launch facility and any space object and give information and assistance to the Launch Safety Officer." [2: 45].

[93] "To ensure compliance with the French Act on Space Operation, a number of civil servants have received the authority to have access, at all times, to the space object and the facilities where the space activity takes place. The operator is notified of such a visit and can observe the visit while being assisted or represented by any person of his choice" [2: 85].

[94] "To enforce adherence to the Act, operators are subject to supervision. The supervisory authority has access to all business premises and facilities, documents and information so as to make sure the supervision is effective" [2: 95].

[95] "To ensure compliance with the provisions of the Outer Space Act, the Minister can request additional information. Furthermore, authorized persons can, at all times, visit the operator's facilities" [2: 110].

[96] Emphasis added.

implementation phases of such activities. In other words, the OST limits itself to establishing the international responsibility of States, in general terms; as a result, the measures that will be chosen to fulfill the obligation of continuous supervision of space activities were left to the discretion of national authorities.

In the same vein, the Sofia Guidelines stipulate in draft Art. 5 entitled *Supervision* that "All space activities are subject to *continuing supervision by the ministerial authority*. Details of such shall be laid down in an implementing decree/regulation.[97]". Yet, to facilitate the alignment of domestic legislation with international space law, the comment written under draft Art. 5 clarifies that supervision of space activity "shall be entrusted to the same authority responsible for the licensing. Furthermore, it should be made sure that such information is available on the governmental level".

This last requirement—namely, to ensure that "such information is available on the governmental level"—should be seen in connection with Art. VI of the OST, stating that no space activity is detached from any governmental responsibility. In other words, the view under draft Art. 5 elaborates on the obligation that all types of space activities are subject to "authorization and *continuing supervision by the appropriate State Party* to the Treaty[98]", and provides that States must be given all the (administrative and technical) means to exercise such supervision.

On this basis, Art. 8 of law n° 4508/2017 laid down rules and procedures concerning the supervision and control of space activities, and essentially established the right of the competent authority to control—at any time, after the granting of the authorization to conduct a space activity—all the installations and materials used by the operator, with the assistance of experts when required. More precisely, Art. 8.1 created a specific right of inquiry to the benefit of said authority: the Minister of Digital Governance may appoint experts—natural or legal persons, whether governed by public or private law—, having a several-years international experience on space matters and/or on aerospace systems, for the purpose of providing them with assistance in supervising the activities carried out by the operator. The type of assistance provided by experts to the supervising authority is not specified and/or limited in any manner.

Correspondingly, each operator is bound to take all measures necessary to facilitate the conduct of on-site controls and inspections of any space activity carried out, and at any time [1: Art. 8.1]. It follows that, for the purpose of such inspections and controls, experts have a right of access to all the information and materials belonging to the space operator, as well as to updated information and data resulting from space activities. At the same time, they have a right of access to installations and infrastructure used (either directly or indirectly) to carry out the space activity [1: Art. 8.2]. As a result, Art. 8.1 and Art. 8.2 of the Greek space law provide the competent authority with enough powers to conduct a in-depth inquiry, assisted by

[97] Emphasis added.

[98] OST, Art. VI (emphasis added).

experts—that may control such issues as technical matters, progress made or risk factors—while enjoying extended investigatory authority.

On the other hand—as mentioned above—the Greek legislator took into account that the space sector is highly based on secrecy and competitiveness. Hence, to balance such broad powers of investigation, both the competent authority and the experts are required to keep any information gathered and supplied by the space operator, secret. Literally, Art. 8.3 stipulates that all the data obtained by experts, during their mission to control the space operator and the space activity, must be treated with confidentiality [1: Art. 8.3].

Once more, the space operator's refusal to grant the access requested by experts can be considered as sufficient grounds for suspending or withdrawing the authorization to conduct a space activity that has been granted [1: Art. 8.4]. Interestingly, apart from what is mentioned in Art. 8.4, no other provision of the law sets out how the outcome of such controls should be assessed. Therefore, it is reasonable to argue that the supervising authority shall be free to decide whether and/or what measures should be taken, when such an investigation ends; the competent minister will, in this sense, be able to decide independently and essentially free from any other influence, either such influence is exercised by other State authorities or by any entities (or persons) outside the administration.

5.2.3 *Withdrawal, Suspension, and Amendment of Authorization*

States must guarantee that all space operations implemented under their jurisdiction are conducted in accordance with the requirements of the authorization granted to the space operator(s), and in line with international space law. Indeed—as it was outlined above—Art. VI of the OST is binding upon States as to the result to be achieved (i.e., as regards the "authorization and *continuing supervision* by the appropriate State Party to the Treaty[99]"), while leaving domestic authorities the choice of the most appropriate forms and methods in view of its detailed implementation. For this purpose, national legislators set *ab initio* specific authorization conditions for space activity; coupled with that, arrangements are usually made to allow the competent authority be promptly informed of any significant failure of the license holder to comply with the conditions set out in the authorization.

Put differently, in case of a failure to comply with the conditions of the authorization—or for any other significant reason occurring in the course of a space activity—, legislations tend to provide that the authorization which was provided may be amended, suspended or withdrawn. "Simply put, the license can be modified, revoked or suspended if the licensee does not comply with the basic requirements, that is, that the activity impairs public health, the national security of

[99] Emphasis added.

the state and the national interest of the state, breaches an international obligation of the states or does not comply with the conditions set out in the license, regulations or Act" [2: 155].

On this basis, it was suggested—in the Sofia Guidelines—that measures allowing the competent authority to supervise and assess the safety of each space activity should be established in national laws [7: Art. 6], without however any prejudice to other actions that States could take against the defaulting license holder. Following this reasoning, draft Art. 6 stipulates that "The respective authority may withdraw, suspend or amend the authorization, when either the conditions of article 4, paragraph 1,[100] or the specific requirements of article 4, paragraph 3,[101] are not observed.". As a result, the competent authority may decide on and implement the authorization (which may be unilaterally withdrawn, suspended or modified, in case the conditions for granting it are no longer fulfilled) and therefore has the final say.

In reference to the Greek space law, Art. 9.1 reaffirms this approach, and lays down that the Minister of Digital Governance may withdraw or suspend the authorization to carry out a space activity, in one of the three cases explicitly provided for: namely, in case that (i) the general or particular conditions of the authorization are no longer complied with [1: 9.1.(α)]; (ii) in the event of an infringement of the provisions on the granting of a space activity authorization [1: 9.1.(β)] and (iii) for reasons of public order, security and health [1: 9.1.(γ)].

Following on from that, Art. 9.2 clearly stipulates that prior to the adoption of such a decision by the competent authority (i.e., in case of a withdrawal or suspension of an authorization *which has already been granted*), the space operator shall be notified of the possibility to provide clarifications and adopt any required measures to remedy the situation. In this context, the space operator may be given the opportunity to provide explanations, and suggest a number of measures to resolve existing infringement(s) of the law or address other violations, within a specific time limit to be decided upon by the competent minister. Importantly though, it is additionally established that in the event of an emergency, the authorization may be revoked without any prior hearing of the space operator .

In essence, Art. 9.2 laid down the right to a prior hearing and the right of defense of the operator (i.e., the licensee): indeed, the law gives them the opportunity to make their views and objections known, before any final decision of the supervisory authority—i.e., adversely affecting the conduct of the space activity—is adopted and implemented. However, this right must be exercised within a reasonable period of

[100] Draft Art. 4.1 mentions all the conditions required to grant an authorization, namely: (a, b) that the operator must be in a financial position to undertake space activities, as well as be reliable and with all the required technical knowledge; (c, d, e, f, g) that the space activity must not cause environmental damage, must be undertaken in such a manner as to mitigate any potential space debris and must be compatible with public safety standards, in line with national security interests, international obligations and foreign policy interests and (h, i) that the operator must have complied with all ITU Regulations and insurance requirements.

[101] Draft Art. 4.3 contains a general clause: "The authorization may contain conditions and requirements".

time, which shall be determined by the competent minister. The fact that the authority is not bound to apply the rules on the prior hearing of the space operator (s) in the event of an emergency [see 1: Art. 9.2] is explained by the necessity to guarantee a rapid—and in this sense, efficient—intervention in such a case, taking into account the difficult context of space as well as all interests at stake.

Nonetheless, it is also true that, from an overall practical perspective, any unexpected withdrawal, suspension or amendment of a granted authorization may have adverse effects *first* on the financial interests of the parties involved, thereby putting at risk the conduct of the space operation(s). In light of this, the legislator took measures to ensure the satisfactory completion of the ongoing activities, *via* the adoption of a transition plan for their safe management until such completion, taking into account *first* the need to assist the space operator(s) in their effort to fulfill their contractual obligations. Thus, Art. 9.3 of law n° 4508/2017 stipulates that, in case of a withdrawal or suspension of an authorization, the Minister of Digital Governance may adopt any provisional measures deemed necessary to ensure the appropriate management of the space activities initiated by the space operator under the authorization regime in question, but not yet concluded, to assist them in meeting their contractual obligations. However, to benefit from this kind of measures, the operator has to address a formal request to the competent authority, without undue delay, immediately on becoming aware of the authority's decision to withdraw or suspend the authorization [1: Art. 9.3].

Be that as it may, the withdrawal—or the suspension—of an authorization to conduct a space activity can (*second*) pose technical problems at the level of its implementation (due to the complexity of space operations); especially given the additional expenses that may be caused by the unplanned termination of said operations. In truth, such aspects (namely, unforeseen events and complications) must always be taken into consideration, and any decision having an impact on the flight operation or guidance of a space object must be informed by best practices in the field, in accordance with safety standards and the methods permitted. In this regard, law n° 4508/2017 provided for the adoption of any required measures, with regard to the need to also ensure the safety of persons and assets, and the protection of the environment.

Hence, Art. 9.4 mentions that the competent authority may order the space operator—as well as any third party—to take all the necessary measures needed to ensure the temporary continuation, or the safe termination of the space activity, and also guarantee the safety of persons, as well as the protection of assets and the environment. In order to achieve this, the competent authority may use the services provided by third parties at their discretion, and/or decide on the transfer of the space activities concerned to a different space operator so as to guarantee the continuity of operations (such as e.g., by ensuring the control and navigation of the space object). Eventually, they may—if necessary—decide to order the removal (de-orbiting), de-activation or destruction of the space object, for the same reasons [1: art. 9.4];

importantly, comparable rules have also been adopted in other legal systems.[102] In all cases, the ministerial decisions—on the suspension and/or the withdrawal of the authorization to carry out a space activity—will be published in the Greek Government Gazette [1: Art. 9.5].

As could have been expected, there were numerous reactions—within the context of the public consultation on the draft of the space legislation [13: Art. 9]—to the adoption of the last case mentioned in Art. 9.4 (i.e., that the competent authority may also order to "destroy the space object"), given that the destruction of a space object (in the traditional sense) will create debris; which will pose additional risks to the environment of space itself, as well as to the other space activities. The same wording was maintained in the final version of the Greek space law; still, it remains to be seen how the term *destroy* will be interpreted and concretized (especially, taking into account the explicit rule in Art. 4.1.(δ) laying down that the space operator must have provided for debris mitigation measures on a voluntary basis[103]). Hence, the likely approach to Art. 9.4 is that any measures adopted by the competent authority must also aim at avoiding the creation of additional space debris, and that any destruction of a space object should be regarded as meaning e.g. its removal, deactivation or de-orbitation. At the same time, one must agree that the rules adopted in Art. 9.3 and 9.4 did create—in general, but also coupled with Art. 16 of the Greek space law laying down transitional measures—an adequate framework to efficiently regulate and address any challenging space activities before they become a real issue.

5.2.4 *Transfer of Space Activities*

Transfers of ownership or control of a space object in orbit are becoming common practice. Usually, such agreements are concluded under specific circumstances to avoid economic loss (i.e., the complexity and specific features of space operations and conditions create a whole set of uncertainties for the space industry, while they at the same time require rapid responses from the market operators).

[102] For instance, according to the UK Outer Space Act, "(W)henever an activity of a licensee is being carried on in violation of the license requirements or the conditions stipulated in the license, directions may be given to ensure that the activities of the licensee adhere to the international obligations of the UK or the conditions of the license (...). In particular, such instructions or directions may include the cessation of the activity or the disposal of any space object. Furthermore, a warrant may be issued that authorizes a named person acting on behalf of the state to do anything necessary, including using reasonable force if necessary, to secure compliance with the international obligations of the UK or with the conditions of the license" [2: 30].

[103] As developed above, law n° 4508/2017, Art. 4.1.(δ), precisely mentions that in order to be authorized: "appropriate provision must have been made for space debris or residues mitigation and management *according to the technological state of the art and international practices*" (emphasis added).

From a legal perspective, the Outer Space Treaty does not prohibit such transfers of ownership; nonetheless, it does not provide for any explicit rule or guidance regarding the transaction in question, either.[104] In reality, the question which practically arises is whether a space object that has changed hands will still fall under the responsibility of the State which initially granted the authorization to use it i.e., within the context of a specific space activity and by (a) particular space operator (s), and undertook to ensure the *continuous* supervision of its uses and operation [34: 29 et seq.].

Literally, Art. VI of the OST stipulates that all space activities, including those of non-governmental entities, require authorization and *continuous supervision by the appropriate State*. In practice, the term *appropriate State* is not clearly defined in the Treaty,[105] but it may be regarded as applying to the State of registry, taking essentially into account Art. VIII of the OST which mentions that the State of registry[106] is the one that shall retain jurisdiction and control over the space objects listed in its system. In this context, although the State of registry (which is one of the launching states, in line with Art. I of the Registration Convention, as mentioned above) retains jurisdiction and control over a space object—on the basis of Art. VIII of the OST—, it will lose its real and effective control over the space object (i.e., in terms of operational control[107]) in case it is transferred. At the same time, the new owner will acquire control over it, but will not be able to obtain jurisdiction by way of registration, unless it was an original launching State for that space object.[108] Hence, the end result of this paradox is that, even though the space object has been transferred (and is now under the operational control of an entity under the

[104] "The concept of 'transfer of ownership' is completely absent from the major space treaties referred to earlier (. . .). Also, the ITU Constitution, ITU Convention and ITU Radio Regulations, the main triad of ITU treaty documents providing another set of legal instruments indispensable for space activities in view of the need for interference-free usage of certain radio frequencies, do not contain a single reference to the concept of 'ownership', let alone of 'transfer of ownership'" [34: 31–33].

[105] "The term 'appropriate State' has concerned many scholars (. . .). In general terms it has been argued that the State of nationality or the State responsible for the activity or the launching State or any State involved to the launch could qualify as the "appropriate State". However, since Article VI sets no other criterion than the mere act of authorization and supervision of a space activity, it suffices to spot the State best qualified best to do so" [35].

[106] Which is one of the launching States, according to Art. I.(c) of the Registration Convention: "The term 'State of registry' *means a launching State* on whose registry a space object is carried in accordance with article II." (emphasis added).

[107] "(T)he (governmental or not governmental) entity asserting control of a space object practically has immediate and direct influence over it, as the sole entity in a position to programme, manoeuvre and operate it in any way." [35].

[108] "(. . .) a State may in theory exercise quasi-territorial jurisdiction over the satellite, but in case of an international transfer of ownership may actually see such a possibility slip out of sight, whilst the 'new' State of nationality of the new owner may not have such opportunities to obtain jurisdiction by way of registration if it was not at least an original launching State of that space object itself" [34: 36].

jurisdiction of a different State), the (initial) State of registry could still be held liable on the basis of Art. VI of the OST.

Some States, like the Netherlands, expressed the view that transfers of authorization may be prohibited for security reasons, like "to prevent the emergence of a market wherein entities emerge that are specialized in adhering to the requirements and conditions of the Dutch Space Activities Act, which allows them to then sell the license to other entities that do not adhere to the requirements and conditions" [2: 155]. Following this reasoning, a few countries established the rule that an authorization cannot be transferred by one space operator to another.[109] Still, it can be argued that solutions can be found to meet the supervisory concerns of States while taking due consideration of structural changes in the space sector, such as the increased privatization and commercialization resulting in satellites being sold and bought while already in orbit.[110]

In view of these concerns, the Sofia Guidelines clarified that "(a)ny transfer of space activity to another operator *may cause additional problems*. It must be secured that the new operator meets the same conditions for conducting the respective space activity (. . .) any such transfer of a space activity should have the authorization of the competent authority (. . .)[111]". Therefore, on this specific basis, draft Art. 9 stipulates that "(t)he transfer of a space activity and/or a space object to another operator is subject to prior authorization by the competent authority. Authorization will be granted under the conditions laid down in article 4[112]".

In the Greek space law, Art. 10.1 laid down the principle that any intended transfer of a space activity to a third party is *subject to prior authorization* granted by the competent authority, today the Minister of Digital Governance (as for a request for a space activity authorization). In particular, any request for a space activity transfer must be presented *jointly* by the transferor—i.e., the actual holder of the space activity authorization—and the transferee (namely, the transfer beneficiary) [1: 10.2]. The conditions and rules under which an authorization to carry out space activity is granted shall apply similarly, in the event of a transfer of the space activity to a new operator [1: 10.3]. Interestingly, the Greek space law underlines that the Minister of Digital Governance is empowered to impose any additional terms and conditions they consider necessary under the new circumstances, either on the transferor or on the transferee, on occasion of the transfer of the space activity [1: Art. 10.4].

Finally, the Explanatory Memorandum to the draft of law n° 4508/2017 underlined that the aim of the legislation is also to ensure that the Greek State is

---

[109] This is the case, *inter alia*, for China, Belgium and the Netherlands [2: 51, 59 and 70].

[110] "This can also be prevented by regulating the transfer as the other states have done. In such a case, the Netherlands can prevent the emergence of a market by not giving the authorization of the transfer of the licence if it suspects that the original licensee is an entity that is specialized in selling licenses" [2: 155].

[111] Emphasis added; see Sofia Guidelines, comment under draft Art. 9 entitled *Transfer of space activity* [7].

[112] Art. 4 of the Sofia Guidelines regulates the conditions required for authorization.

protected in the event of a claim being brought against it on the grounds of international space law, in particular with regard to responsibility for, and supervision of space activities and space objects. Hence, the Greek space law precisely provided for cases where the transferee—i.e., the beneficiary of the activity transfer—is a foreign natural or legal person [3: 3, comment under Art. 10], due to the critical issue of the remaining responsibility of the State of registry even after the ownership of the space object is transferred.

In substance, this point was regulated by Art. 10.5, clarifying that such transfer shall be authorized only if the country in question—i.e., the country of the beneficiary of the transfer—genuinely provides for an adequate level of protection for the interests of the State of the transferor. In other words, the Greek Minister of Digital Governance is entitled to refuse authorization of a space activity transfer, if there is no agreement with the transferee's country which would allow the Greek State to recover compensation, in case Greece has paid damages on the basis of its international liability under Art. VI and VII of the OST (i.e., for damage caused by the space object that changed hands).

5.3 Risks and Hazards in the Context of Space Activity

It is common knowledge that space activities are inherently challenging to organize and manage, due to the high costs involved and the unknown effects that the risky space environment may have on both the human body [36] and space assets. At the same time, space appears to be a highly promising market and the related economy is witnessing an unprecedented growth. As a result, there is a constant increase in the conduct of space activities resulting in the proliferation of objects being used for the exploration and use of space resources; which growing number *ipso facto* generates more threats, due to increased chances of having a collision and creating damage(s) to other operating objects (used either for civilian or for military purposes; as civilian, commercial and military activities remain—up to now—closely intertwined in space) [37]. Hence, special attention had to be given to the issue of space threats as well as to the necessity to ensure the safety of space objects and activities, in general.

From a technical perspective, several types of damage—or risks of damage—from space activity may arise when launching the space object (in this case, damage can take the form of explosive, toxic, or radioactive hazards[113]). In a next stage, damage can be caused by a collision either during proximity operations (i.e., rendezvous and docking) or with other space traffic [38: 226]; the latter being increasingly probable due to the growing amount of space debris, coupled with the

[113] "As of today, there have been nearly 200 people killed on ground by rocket explosions during processing, launch preparations, and launch" [38: 207].

fact that the survival time of said items may be particularly long.[114] Eventually, the damage could also be the direct result of a space object's reentry into the atmosphere [38: 216].

The Outer Space Treaty does not distinguish between natural and man-made threats; it merely establishes a general obligation to report *any phenomena* discovered in outer space, in case these may be regarded as potentially dangerous for the life or health of astronauts (see OST, Art. V para. 3). Interestingly, apart from that specific provision and notwithstanding the importance of the issues caused by space-based threats,[115] "no international regulation applicable worldwide has been agreed" yet [38: 220]. Hence, to enhance legal certainty for space operators, States have adopted space situational awareness systems (SSA)—to monitor the risks caused by objects in orbit, natural space objects approaching the Earth (Near-Earth Objects/NEO) etc.[116]—, in addition to more specific legislation aimed at encouraging operators to adopt measures and develop practices contributing to reducing risks related to their activities.

By way of illustration, in order to deter space operators from engaging in unsafe and/or illegal behavior likely to result in the creation of additional risk factors in space, Indonesia established that "the launching of space objects that intentionally does not comply with the requirements and that causes damage to goods or persons shall be punished with a maximum of 15 years imprisonment or a fine of maximum Rp 4 trillion (approximately €250 million). The sanction is increased to a maximum of 20 years imprisonment or a fine of Rp 5 quadrillion (approximately €320 million) if loss of life is caused by the accident" [2: 102]. Correspondingly, the Indonesian Space Act also laid down that, should any kind of calamity occur, "the license can be revoked or amended as is deemed necessary during the suspension period" [2: 102].

As regards the Greek space legislation, risks are first addressed on an *ex-ante basis*, that is with the aim to essentially prevent and avoid hazards; the measures concerned mainly consist of the obligation to duly inform competent authorities about any dangers that may occur in space, coupled with specific actions for their avoidance. Along with that, the Greek legislator adopted rules governing liability and compensation in those cases where a risk actually materializes, following an *ex-post evaluation* of the damage.

[114] "Scientists assume that there are approximately 500,000 objects in orbit whose sizes are above 1 cm. (…). Objects in 1000 km orbits can exist for hundreds of years. At 1500 km, the lifetime can go up to thousands of years. Objects in geosynchronous orbit can presumably survive for one million years", [38: 211].

[115] For instance, "Loss to civil and commercial customers is measured in financial terms; while loss to the military is measured in operational terms" [37: 50].

[116] See, *inter alia*, the European SSA, an essential component of the EU Space Programme [39].

5.3.1 *Risks and Threats for Space Activities:* Ex-ante Measures

To address the threats and hazards that may occur in outer space, international space law lays down minimum preventive measures of a rather general content, mainly aimed at increasing the awareness and preparedness of the space stakeholders. More precisely, an obligation to inform the competent national and international bodies was established, in case danger—requiring urgent response action— materializes. Furthermore, taking into account that space constitutes a particular environment presenting numerous risks and uncertainties which are sometimes difficult to predict and address, cooperation was adopted as a fundamental element of this framework.[117] As a result, a general obligation to stay alert and take all the necessary and appropriate measures (and protect the safety of all) lies on all States i.e., including Greece, and must therefore be regarded as a matter of primary importance by all the space operators conducting space activities under the jurisdiction of the Greek State.

5.3.1.1 The International Law Obligations of States to Inform and Cooperate

First, it is necessary to note that international space law established the unequivocal obligation of all States to inform—other States, as well as the competent international authority—of risks that may arise in the context of space. Indeed, focusing on a specific aspect of space activities (i.e., the necessity to ensure the protection of astronauts and all personnel onboard), the OST laid down in Art. V para. 3 the general duty of States to inform in case of "any phenomena they discover in outer space, including the Moon and other celestial bodies, which could constitute a *danger to the life or health of astronauts*", as mentioned above (emphasis added).

Later, as was reasonably expected, Art. 1 of the Rescue Agreement clarified this rule, by mentioning that any State which "receives information or discovers that the personnel of a spacecraft have suffered *accident* or are experiencing *conditions of distress* or have made an emergency or unintended landing in territory under its jurisdiction or on the high seas or in any other place not under the jurisdiction of any State (...)[118]" must notify the launching authority as well as the Secretary-General

[117] Indeed, "Cooperation is (...) stipulated in different forms, in Articles I, III and IX of the Outer Space Treaty", [2: 183].

[118] Emphasis added; in this case, the State party which receives such information shall: "(a) Notify the launching authority or, if it cannot identify and immediately communicate with the launching authority, immediately make a public announcement by all appropriate means of communication at its disposal; (b) Notify the Secretary-General of the United Nations, who should disseminate the information without delay by all appropriate means of communication at his disposal", see Rescue Agreement, Art. 1 para. (a) and (b).

of the United Nations. The obligation to inform is also further elaborated in other articles of the Rescue Agreement, such as in Art. 2[119] and Art. 3.[120] In this context, for instance, a duty to "furnish information as soon as feasible [*i.e., in cases where a space object is involved in an incident, calamity or disaster*] exists under the Dutch Space Activities Act (...). The Netherlands considers the provisions necessary in order to adhere to its treaty obligations to cooperate with other states that might be affected by the incident" [2: 182–183].

Furthermore, a general duty to provide information on the nature, conduct, locations and results of space activities—in addition to the characteristics of any operating space objects—was laid down in both the OST[121] and the Registration Convention, for a preventive purpose. In reference to the latter, Art. IV of the Registration Convention precisely enumerates the pieces of information that must be provided by States to the Secretary-General of the United Nations[122] and updated on a regular basis,[123] to keep the international community informed about these objects overall operation and use (* *this obligation of States to register space objects may be regarded as a broad duty aimed at improving space traffic safety and management 'per se'. In other words, registration is not always related to risks and liability issues; it is first a general and preventive measure aimed at being able to identify the space object in many cases until the (safe) termination of the activity. It is therefore presented as such here, that is as a necessary and important practice which is not always revolving around the idea of risks materializing in space*[124]).

Interestingly, no specific provision—precisely devoted to the adoption of preventive measures (such as for information dissemination)—is contained in the Sofia

---

[119] Rescue Agreement, Art. 2: "If, owing to accident, distress, emergency or unintended landing, the personnel of a spacecraft land in territory under the jurisdiction of a Contracting Party, it shall immediately take all possible steps to rescue them and render them all necessary assistance. It shall inform the launching authority and also the Secretary-General of the United Nations of the steps it is taking and of their progress (...)".

[120] Rescue Agreement, Art. 3 applies to the personnel of a "spacecraft [*that*] have alighted on the high seas or in any other place not under the jurisdiction of any State".

[121] OST, Art. XI: "In order to promote international cooperation in the peaceful exploration and use of outer space, States Parties to the Treaty conducting activities in outer space, including the Moon and other celestial bodies, *agree to inform the Secretary-General of the United Nations as well as the public and the international scientific community, to the greatest extent feasible and practicable*, of the nature, conduct, locations and results of such activities", (emphasis added).

[122] The Registration Convention enumerates, in Art. IV para. 1, the following information: (a) Name of launching State or States; (b) An appropriate designator of the space object or its registration number; (c) Date and territory or location of launch, as well as specific technical characteristics and orbital parameters, such as (d. i) Nodal period; (d. ii) Inclination; (d. iii) Apogee (d. iv) Perigee; and (e) General function of the space object.

[123] See, Registration Convention, Art. IV para. 2: "Each State of registry may, from time to time, provide the Secretary-General of the United Nations with additional information concerning a space object carried in its registry".

[124] The obligation to register space objects is thoroughly examined below under Sect. 5.4.1.

Guidelines. In this sense, the instrument seems to reflect the view that the issue of threats in outer space could be tackled using a combination of the existing obligations and incentives; which could include the duty to ensure the supervision of space activities[125] and the obligation to provide for environmental protection.[126] Conversely, the Guidelines did put emphasis on remediation measures.

5.3.1.2 National Level: Obligation of the Space Operators to Inform the Greek State

Following on from the above, it appears that—according to international space law—States have the obligation to provide information on (more or less) *any threat* arising in the context of space activities, whether these are carried out by governmental agencies or by non-governmental entities. In other words, States must communicate information on activities under their jurisdiction (and on risks materializing in this context), to the other States and to the competent international authority, irrespective of the legal status of the operator (as a corollary to the rule established in Art. VI of the OST). Therefore, the national legislator had to ensure that the State will be able to effectively receive this information in cases where space operations are carried out by non-State entities, namely by private space operators; especially when legal persons constituted under foreign law are being involved.

On this basis, Art. 12 of the Greek space law imposes on any operator an extended obligation to inform *without any delay* the competent authority about *any* problem (s) encountered in the context of space activity. The underlying aim was to ensure that the Greek State is given the means to respond as quickly as possible to any potential risks materializing in space, using all reasonable and available steps to prevent injury to persons as well as any loss or damage caused to property and/or to the environment.

More specifically, Art. 12 lays down two obligations regarding the information that must be forwarded to the competent Minister (i.e., the Minister of Digital Governance). *First*, the operator must communicate any problem of a technical or operational nature, namely any issue regarding the command of the object, as well as any malfunction or abnormality detected; in addition to that, the operator must inform of any general loss of control over the space object which could eventually put at risk people on Earth, airplanes in flight or any other space objects, or cause any type of damage. *Second,* the operator must also notify the competent authority of any changes in the administrative information that has been provided, to allow the Greek State to have an accurate understanding of the object and the activity being conducted, throughout the whole operation and until its termination; namely, the

[125] Addressed in the Sofia Guidelines, in draft Art. 5 [7: Art. 5].

[126] Also addressed in the Sofia Guidelines, in draft Art. 7 entitled *Protection of the environment*, coupled with draft Art. 8 on the *Mitigation of space debris*.

operator must notify "*any* change related to the information contained in the application for authorization", (emphasis added) [1: Art. 12].

Finally, as the purpose of the Greek legislator was to address *all* potential sources of risks in space, attention was also given to regulating particular kinds of situations that may be regarded as being—from a certain perspective—*inherently riskier*, such as (i) unauthorized space activities and (ii) activities that had been started at a time when there were no national measures in force, namely space operations already underway when the Greek space law was adopted.

5.3.1.3 The Rules Applying to "Inherently Riskier" Situations

Given the rule established in Art. VI of the OST (i.e., that States bear international responsibility for *all* their activities in outer space), Greece had to give specific attention to activities which—for some reason—have not been, or had not been, initiated following a proper control and authorization procedure conducted by the competent space authority. In line with this reasoning, the Greek space law devoted two articles to regulate activities that could be considered "inherently risky" situations: unauthorized outer space activities [1: Art. 15], and activities that had been started before its adoption and were still ongoing [1: Art. 16].

5.3.1.3.1 Art. 15: Addressing the Issue of Unauthorized Space Activity

As mentioned in the Explanatory Memorandum to the draft of the Greek space law, the conduct of unauthorized (or mis-licensed) space activities may significantly affect—i.e., create risks to—the space environment, or to persons and assets in space. In addition to that, such activities may cause harm to the stability of the Greek State's international relations with any nation or organization, as well as to the interests of actors involved in the exploration and use of outer space resources, in general [3: 4, comment under Article 15]. On this basis, sanctions had to be determined for those operators who do not comply with the law and engage in such practices.

In reference to this point, the Sofia Guidelines explicitly suggested that a provision of national law may impose "a minimum sanction in case of space activities and transfer thereof without a licence [*that*] appears reasonable given the seriousness of the offence. The higher limit should be set at a deterring amount similar to breaches of administrative law not reaching, however, the level of a criminal sanction", [7: see comment under draft Art. 14]. Hence, draft Art. 14 established that: "Any breach of the obligations set out in the present law is punishable with a fine of (. . .). The carrying out of space activities (. . .) without authorization from the authority (. . .), is punishable with an amount not lower than (. . .)".

Adopting this view, the national legislator stated, in Art. 15 para. 1 of law n° 4508/2017, that "(a)ny natural person and any duly empowered representative of a legal person, carrying on space activities without an authorization, is liable to punishment of up to two (2) years of imprisonment, and a fine of between 30.000 and 300.000 Euros". Furthermore, if the Greek State were held internationally responsible for the conduct of an unauthorized space activity, it would automatically be entitled to seek legal redress against the offending party—namely, any natural or legal person(s)—carrying out such space activity, and claim compensation; the precise amount of this compensation would be determined in accordance with para. 3 of Art. 11 of the Greek space law, on damage assessment [1: Art. 15 para. 2].

At the same time, the mention of incorrect or incomplete information in any of the documents submitted to the competent licensing authority is punishable under Art. 15 para. 3. More precisely, said provision stipulates that "He [*or she*], who at the application stage and until the authorization expires, intentionally provides inaccurate or incomplete information to the Minister of Digital Policy, Telecommunications and Information (i.e., today, to the Minister of Digital Governance) in reference to said space activities, is punishable by a fine of 10.000 to 100.000 Euros". Interestingly though, the applicant is liable, in this specific case, to a lesser—i.e., compared to the one mentioned above, in Art. 15 para. 1—but still quite significant penalty; the purpose is evidently to also discourage operator(s) from providing the competent authority with false or erroneous information, as this would make it practically unfeasible for said authority to efficiently monitor and supervise the space activity concerned.

Overall, the rules of the Greek space legislation presented above are comparable to measures adopted in other legal systems. Indeed, studies showed that—up to now—most countries have adopted measures of a similar substantive content, aimed at deterring space operators from engaging in unlawful (hence, inherently risky) space activities or related practices. By way of illustration, States such as Sweden mainly set out sanctions in case of breach of the authorization conditions, while others (e.g., Austria, Indonesia, New Zealand) additionally clarify that further measures may also be taken if space activities do not comply with the applicable laws and/or false information has been provided [2: 180–181].

5.3.1.3.2 Art. 16: Transitional Measures Applied to Space Activities in Progress

Apart from unauthorized (or mis-licensed) space activity, particular arrangements were also adopted—in Art. 16—to regulate activities that had started before the adoption of the Greek space law and were still in progress. In that case, account was taken of the fact that, *on the one hand*, all these activities had been started without having prior authorization and/or compliance with the conditions for such authorization as defined and established later in the national space legislation; from this perspective, it could not be assured that said operations did not carry risks. *On the other hand*, the actors involved had already made significant investments in the activities in question, based on long-term planning and partnerships; hence they expected, in turn, that specific measures allowing for a smooth transition to the new regime would be adopted in the Greek space law.

Following on from that, Art. 16 para. 1 laid down that space activities already under way at the time of entry into force of the Greek space legislation could be *prima facie* continued for a period of twelve (12) months, starting from the date of entry into force of law n° 4508/2017; namely, without an authorization being required. Nevertheless, it was also clarified that, within these 12 months, no transfer could take place. In parallel, the space operator was explicitly required to submit—again, within the same duration—to the competent authority all the documentation referred to in Art. 5 on the content of the application dossier, and in Art. 6 on the submission of the environmental impact assessments [1: Art. 16 para. 2]. In other words, practically, the whole procedure aimed at informing the competent authority of all the key aspects of the space activity being conducted had to be followed.

Furthermore, the law stipulated in Art. 16 para. 2 that, in the case of space activities in progress, Art. 15 (i.e., laying down provisions governing penalties for non-compliance with prior authorization conditions) would not apply; indeed, applying penalties in such a context would have been virtually meaningless. As a matter of fact, the legislator adopted a practical approach, and focused instead on how to exercise effective control on ongoing activities. Hence, the operator was merely required to provide information and the documents needed to allow for the verification of the accuracy and veracity of the elements enumerated in Art. 5 para. 2.[127] In essence, the main purpose of the provision was to ensure the smooth termination—or transition to the new regime—for all the parties involved, without taking any shortcuts or bypassing the procedure aimed at informing the competent authority.

Finally, a substantially different regime was established for all space activities being carried out with the involvement of the Greek State, in Art. 16 para. 3. In particular, the activities that had been started based on a specific agreement with the Greek State and that had been granted a proper authorization, were now considered to also be duly authorized under the new legislation. In consequence, they were *de jure* subject to the provisions of the new law, without having to go through any other procedures and/or formalities.

Having established a wide array of rules aimed at addressing risks in space activity on a preventive basis, the Greek legislator additionally adopted measures to be applied after harm has occurred.

[127] As already mentioned, Art. 5 para. 2 of law n° 4508/2017makes reference to information related, *inter alia*, to the full identification of the operator and its legal form; his/her representatives etc.; the space activity, such as its description, guarantees for viability etc.; and regarding measures of compliance with international standards, such as those on safety, environmental protection etc.

5.3.2 *Liability & Redress for Damage from Space Activity:* Ex post Rules

Liability in space is based on Art. VI[128] and Art. VII,[129] coupled with Art. VIII[130] of the OST. All these rules are developed further in the Liability Convention while, in addition to that, guidelines are provided in a number of soft law instruments, to take account for the challenging recent developments in the field of space. As an example, the 2004 Resolution on the Concept of 'Launching State' underlined that States should consider—*inter alia*—the conclusion of agreements in accordance with the Liability Convention with respect to joint launches or cooperation programmes, to facilitate the application of international space law and of the Liability convention in particular with regard to the concept of liability; especially given the increase in "activities carried out jointly by government agencies and non-governmental entities, as well as partnerships formed by non-governmental entities from one or more countries" [21].

In essence, States are liable to pay compensation for any damage(s) caused by their space objects. However, depending on the location where the damage has occurred, two different regimes may apply: if the damage occurred on Earth or to aircraft in flight, the launching State shall be absolutely liable to pay compensation;[131] on the contrary, if the damage took place in space, the launching State will be liable only if the damage is due to its fault or the fault of persons for whom it is responsible.[132] In case of more than one launching States, all the States involved may be held jointly and severally liable for the damage [40: Art. IV et seq].

The rules applying in case of damage occurring in space are precisely laid down in the Liability Convention,[133] to allow for the settlement of international disputes through peaceful means. However, this framework has been criticized for being "State-centric" and poorly adapted to the growing role of private operators in this

[128] OST, Art. VI: "States Parties to the Treaty *shall bear international responsibility* for national activities in outer space, including the Moon and other celestial bodies, whether such activities are carried on by governmental agencies or by non-governmental entities (. . .)", (emphasis added).

[129] OST, Art. VII mentions that each launching State "is *internationally liable for damage* to another State Party to the Treaty or to its natural or juridical persons by such object or its component parts on the Earth, in air space or in outer space, including the Moon and other celestial bodies", (emphasis added).

[130] OST, Art. VII clarifies that the State of registry "*shall retain jurisdiction* and control over such object, and over any personnel thereof, *while in outer space or on a celestial body*", (emphasis added).

[131] Liability Convention, Art. II: "A launching State shall be absolutely liable to pay compensation for damage caused by its space object on the surface of the Earth or to aircraft in flight".

[132] Liability Convention, Art. III: "In the event of damage being caused elsewhere than on the surface of the Earth (. . .) the latter shall be liable only if the damage is due to its fault or the fault of persons for whom it is responsible".

[133] See e.g., the rules providing for the State which suffers damage to present a claim for compensation (to a launching state) etc. [40: Art. VIII et seq.].

new context [41]. In fact, behind this criticism lies the truth that private entities are increasingly attracted to and investing in space (either on their own, or in partnerships with State entities and public law operators), contributing to the significant growth of launches and operations of space objects. Hence, the corollary to such growing use of space was a corresponding rise in the risks posed by operating space objects or by space debris (due to the higher probability of collision between space debris and other space objects) [42].

In response to this issue, common standards and preventive measures were agreed upon, such as the *IADC guidelines* and the *Space Debris Mitigation Guidelines of the Committee on the Peaceful Uses of Outer Space* [43]—as developed above—to avoid the materialization of threats in space and protect the investments of the parties involved. However, bearing in mind that damage may indeed occur, specific additional rules of law were adopted at the national level (as a complement to international space law) to allow for reparation for any loss or damage caused in the context of space activity.[134]

5.3.2.1 Regulating Loss and Damage in Space Activity, in the Greek Space Law

In Greek law, the subject of liability in space is regulated by Art. 11 of the national space legislation which, despite being concise, appears to focus on the substance of the issue and should be read in conjunction with the Explanatory Memorandum. Importantly, the latter instrument mentions that, first of all, any liability resulting from space activity carried out by States—and/or their non-governmental entities—falls within the scope of the Liability Convention (*Convention on International Liability for Damage Caused by Space Objects*, 1972), which was indeed ratified by Greece and transposed into the domestic legal order by means of law n° 563/1977 [3: 3, comment under Art. 11; 43]. However, one must also note that, even if Greece were not a State party to the Liability Convention, it could still be liable under the Outer Space Treaty (OST) or under general international law [2: 158].

In this context, Art. 11 of the Greek space law laid down four fundamental rules that address liability, which are precisely establishing: (i) an insurance obligation; (ii) the right of recourse of the Greek State against the space operator; (iii) rules on the calculation of the indemnification and (iv) the right of recourse of the Greek State against any other launching State(s), as explained below.

[134] There is a wealth of research on liability in space. However, the aim of this study is not to go into further detail on this subject, but rather to examine how the relevant articles of Greek space law fit into the international order.

5.3.2.2 The Obligation to Take Out an Insurance Policy

The Explanatory memorandum to the Greek space law highlights that, according to Art. II of the Liability Convention,[135] responsibility for damage caused by space objects lies with the launching State. Therefore, to guarantee they can cover their liability *vis-à-vis* third parties in case of damage caused during a space activity carried out by private entities under their jurisdiction, States establish in their national space laws that private operators should conclude an insurance contract (providing coverage for potential damage caused by their space objects and/or their space activities). In other words, the "insurance ensures that the private entity will have the financial means to be able to indemnify the state for the compensation it must pay under the aforementioned liability" [2: 158].

In line with this reasoning, the Sofia Guidelines made it clear that if a State is liable (under the Liability Convention or general international law) to pay compensation for damage from space objects, the issue which remains to be addressed—in national laws—is the State's right of recourse against the operator. On this basis, draft Art. 11 of the Sofia Guidelines entitled *Liability and recourse* lays down that "(w)hen XY has paid compensation to third parties for damage caused by a space activity in fulfillment of its international obligations, the Government is entitled to recourse against the operator". Additionally, draft Art. 11.2 mentions that "(t)he recourse of the Government against the operator may be limited to a certain amount" [7: Art. 11].

In this context and as a complement to international law (namely, to the provisions of the OST and the Liability Convention), Art. 11.1 of law n° 4508/2017 clearly stipulates that in order to carry out space activity, the space operator is bound to take out an insurance policy covering their liability for damage, up to the amount of 60.000.000 (sixty million) Euros. That being so, two remarks need to be made before elaborating further on this obligation below.

First, regarding the specific wording used in Art. 11.1 of the Greek space law addressing the question of liability in space, it should be noted that the two different words *responsibility* and *liability* (used alternatively in international space law treaties; e.g., Art. VI of the OST makes reference to the "international *responsibility*" of States whether Art. VII mentions that they are "internationally *liable*" giving rise to concerns about the precise scope of each concept) are translated into the same Greek term: that is, international responsibility [*διεθνή ευθύνη*].[136] Interestingly, the

[135] Which must evidently be read within the context of the fundamental rules established in the OST, such as those concerning the responsibility of States for *all* their space activities, whether carried out by governmental agencies or by non-governmental entities.

[136] Art. VI of the OST mentions that "States Parties to the Treaty shall bear international *responsibility* for national activities in outer space (...)", (emphasis added); which is being translated into Greek as the expression "θα υπέχουν *διεθνή ευθύνην*". Likewise, Art. VII of the OST stipulating that "Each State Party to the Treaty that launches (...), is internationally *liable* for damage (...)", is translated into "Εκαστον συμβαλλόμενον (...) υπέχει *διεθνή ευθύνην* δια τυχόν ζημίαν", (emphasis added); see [44] for the Greek translation of the OST. The same expression [*διεθνή ευθύνη*] is used for the translation of the Liability convention in Greek law, see [45].

same perspective may be found in other legal systems as well,[137] which in a sense allows to practically sidestep the challenging issue of the exact scope and interpretation of the two distinct concepts mentioned above within the specific context of international space law (i.e., responsibility and liability).[138]

Second, apart from a general reference in Art. 2.12 of the Greek space law titled *Definitions* (i.e., according to Art. 2.12, *damage* in the context of space is "any type of damage as defined in the first article of the Liability Convention"), the term *damage* was not re-defined or explained further in Art. 11.1 dedicated to the question of liability in space; such as, for instance, to better take into account any characteristics of the Greek space sector, new technologies used in this context, or any needs of the emerging space industry. Hence, the damage mentioned in Art. 11.1generally refers to any "loss of life, personal injury or other impairment of health; or loss of or damage to property of States or of persons, natural or juridical, or property of international intergovernmental organizations", in line with the definition contained in Art. I.(a) of the Liability Convention.

Following on from the above, the obligation of the space operator to take out an insurance contract covering the *responsibility for damage* that may occur within the context of space activities (established in Art. 11.1) should be interpreted *lato sensu*, and in line with international space law. Be that as it may, the Greek legislator provided for adjustments of this obligation, considering the need to also promote and encourage the participation of non-commercial space operators *inter alia*.

5.3.2.2.1 Adjustments of the Obligation to Take Out Insurance in Greek Law

As already mentioned above (i.e., under Sect. 5.2.1), Art. 4.1.(ζ) sets out that meeting the insurance requirement established in Art. 11 of law n° 4508/2017 is a prerequisite to the granting of authorization to conduct space activity under the jurisdiction of the Greek State.[139] Following on from that, the Greek law provides that *all* space operators are expected to take out liability insurance for an amount adequate to cover the potential damage that could be caused by space objects used in the context of their space activities. More precisely, Art. 11.1 stipulates that space operators must take out an insurance coverage for an amount of 60.000.000 (sixty million) Euros.

[137] "(B)oth the French and the Spanish languages—after all, both authentic languages as far as the Outer Space Treaty is concerned!—have only one term for the two notions ("responsabilité" and "responsabilidad" respectively)" [24: 363].

[138] In fact, liability is "in regard of space law (and in contrast to general international law) the most elaborated of the two principles as a special Liability Convention was devoted to develop the provisions of Article VII of the Outer Space Treaty" [24: 365].

[139] Law n° 4508/2017, Art. 4.1: "The authorization is granted for an undetermined or a determined period and shall be renewable, when the following conditions are fulfilled (...) (ζ): "the operator satisfies the requirements for insurance cover provided for in Article 11".

Nonetheless, it was explicitly suggested (during the public consultation of the Greek space law) that factors such as the number and the financial value of the space objects being used, as well as the economic importance of the space activity concerned, should also be taken into account when calculating the insurance coverage [13: Art. 1]. Thus, Art. 11.1 of law n° 4508/2017 provided, in its final and current version, that the amount of the insurance may as well be adapted—considering several elements such as the size, economic value, and other characteristics or operating conditions of the space object—, after a decision of the competent authority (today, the Minister of Digital Governance) and following a request of the space operator. In such a case, the request to adapt the precise amount of the insurance must be made prior to the date of submission of the application dossier [1: Art. 11.1].[140]

Importantly, in case a space activity is pursued in the broader public interest, the competent authority—i.e., the Minister of Digital Governance—may reduce the amount of insurance coverage, or even release the space operator from the obligation to take out insurance to cover the risks resulting from the space activity in question[141] (in deciding, the minister will be guided by, and take into account, the risks that could indeed result from the space activity concerned). Nonetheless, to avoid any arbitrariness, Art. 11.1 specifies that the space activities regarded as beneficial for the public interest include those aimed at serving "the purpose of national defense and security, science, research or education".[142] Be that as it may, and similarly to other States,[143] it is also clarified that the obligation to take out civil liability insurance does not apply when the operator of space activities is the Greek State.

5.3.2.3 Right of Recourse of the Greek State Against the Space Operator

Art. 11.2 of law n° 4508/2017 clarifies that in case the Greek State is liable to pay compensation, it shall also be entitled to take recourse against any operator involved in the launching and/or operating of the space object(s) that caused the damage (i.e., for damage caused during a space activity conducted under the Greek State's

[140] Art. 11.1 precisely mentions that such request must be made before the submission of the documentation referred to in law n° 4508/2017, Art. 5.2. (Art. 5.2 enumerates the information that must be presented with each application for authorization of a space activity).

[141] Similar rules were adopted in other countries. For instance, according to the Austrian Outer Space Act, some activities can be "exempted from having insurance or have the required insurance limited by administrative decisions. However, this applies only if the space activity is conducted in the public interest, meaning the activity serves science, research or education." [2: 91].

[142] The Greek law seems to establish an exhaustive list of cases, as it refers that: "Space activities are in the public interest *if they serve* the purpose of national defense and security, science, research *or* education", (emphasis added) [1: Art. 11.1].

[143] *Inter alia*, see the Indonesian space act: "the operator is obliged to provide financial guarantees or take out insurance for launching activities (...). The insurance policy does not apply to governmental entities" [2: 99].

jurisdiction). Therefore, first of all, the State's liability will be determined in accordance with Art. VII of the OST and the rules established in the Liability Convention, as is precisely mentioned in the Greek space law.[144]

More precisely, Greece will be held responsible for any damage caused by a space object, if it is its launching State.[145] On this basis, and as mentioned above, it shall be: (a) absolutely liable for the damages enumerated in Art. II of the Liability Convention ("A launching State shall be absolutely liable to pay compensation for damage caused by its space object *on the surface of the Earth or to aircraft in flight.*[146]"). Otherwise, (b) it shall be held liable for fault in accordance with Art. III of the Liability Convention, laying down that: "In the event of *damage being caused elsewhere* than on the surface of the Earth to a space object of one launching State or to persons or property on board such a space object by a space object of another launching State, the latter shall be liable *only if* the damage is due to its fault or the fault of persons for whom it is responsible[147]" (taking into account, in both cases, that no cap on the State's liability was laid down in the Greek space law[148]). In addition to that, all the rules established in the Liability Convention—such as in reference to the joint liability of States (Art. IV); exoneration from absolute liability (Art. VI); the right to claim compensation (Art. VIII); the right to establish a Claims Commission (Art. XIV) etc.—will apply, where appropriate.

In this context, each time damage is caused by a space object under any operator's responsibility, the Greek State is entitled to exercise recourse against the operator concerned, as provided for in Art. 11.2 of law n° 4508/2017 (other States have also adopted that type of rule, under their national space laws, to create a legal basis allowing them recover the compensation paid for damage caused by objects and activities under their jurisdiction[149]). However, to provide some legal certainty to the space industry, the Greek legislator additionally established a limit to this right of recourse; namely, it was explicitly mentioned in law n° 4508/2017 that, when the State pays compensation and exercises its right of recourse against the space operator (s) involved, it will be entitled to do so only up to the amount set out in line with Art. 11.3 of the Greek space law.

---

[144] Indeed, the Greek space law mentions: "When the Greek State is internationally liable to pay compensation, pursuant to Article VII of the Outer Space Treaty, the provisions of the Convention on International Liability for Damage caused by Space Objects or this law (...)" [1: Art. 11.2].

[145] Given that, according to the OST, Art. VII "Each State Party to the Treaty that *launches or procures the launching* of an object into outer space (...) and each State Party *from whose territory or facility an object is launched*, is internationally liable for damage to another State Party to the Treaty or to its natural or juridical persons by such object or its component parts on the Earth, in air space or in outer space, including the Moon and other celestial bodies", (emphasis added).

[146] Emphasis added.

[147] Emphasis added.

[148] Contrary to that, other States, like Australia, established for instance liability periods (i.e., "liability under the Act seemingly does not extend to damage that is caused, while the space object is in-orbit after the liability period for the launch has lapsed") [2: 43].

[149] Such as, for instance, the Dutch Space Activities Act [2: 73].

Needless to say, not all States have adopted such a limit; by way of illustration, "the [Swedish Act on Space Activities] does not explicitly limit the amount of compensation for which Sweden can seek recourse. Therefore, the amount of compensation that can be recovered is the full compensation paid by Sweden[150]".

5.3.2.4 Rules on the Calculation of Compensation

In setting out rules for the assessment of compensation for damage, Art. 11.3 of law n° 4508/2017 distinguishes between two different situations, depending on whether the person and/or legal entity that suffered a loss or damage—as a result of the conduct of a space activity under the jurisdiction of the Greek Sate—is (*first*) a third State or a third-country national, or (*second*) a Greek national.

The *first* case is regulated by Art. 11.3.(α), clarifying that when damage is caused to a third State or to third-country nationals, its assessment shall be made through a joint procedure. More precisely, the amount of damage suffered will be debated jointly by the Greek State and the State suffering the damage or representing the victim, in accordance with the specific procedure laid down in the Liability Convention. Art. 11.3.(α) of the Greek space law does not precisely mention which articles of the Liability Convention should apply in this instance. Be that as it may, it is reasonable to consider that the Greek legislator had—at least—in mind the rules governing the right to claim damages before the competent authorities;[151] the rules on reparation;[152] and the right of parties to request the establishment of a Claims Commission.[153] Finally, Art. 11.3.(α) clarifies that the third-country space operator—or any other person that will be designated by them for this specific purpose—may participate in the debates and damage assessment procedures, together with the representatives of the State(s) involved, in order to defend their interests.

The *second* case, referring to damage being caused by any space object (s) operating under the Greek jurisdiction to Greek nationals, is regulated in detail in Art. 11.3.(β). In particular, it stipulates that in case of damage caused to Greek nationals, the assessment of damage shall be conducted by a committee consisting of three (3) experts. Each party to the dispute shall be given the right to appoint one member of the expert committee, while the third member of the expert committee will be appointed by a common agreement between the parties. Interestingly, the legislator established that the competent minister—today, the Minister of Digital Governance—may also decide that such experts must be designated *ex-ante* as one

[150]"Consequently, having the financial ability to reimburse Sweden can be one of the restrictions or conditions to be granted the license" [2: 20].

[151]Liability Convention, Art. VIII et seq.

[152]Liability Convention, Art. XII et seq.

[153]Liability Convention, Art. XIV et seq.

of the conditions for issuing authorization to carry out space activity [1: Art. 11.3.(β)].

5.3.2.5 Right of Recourse Against Other (i.e., Co-launching) States

Finally, Art. 11.4 of the Greek space law made specific reference to the State's right to recourse against any other co-launching State, in case of a space activity where more than one launching State is involved. In fact, said article explicitly references Art. V para. 2 of the Liability Convention, establishing that "(a) launching State which has paid compensation for damage shall have the right to present a claim for indemnification to other participants in the joint launching. The participants in a joint launching may conclude agreements regarding the apportioning among themselves of the financial obligation in respect of which they are jointly and severally liable. Such agreements shall be without prejudice to the right of a State sustaining damage to seek the entire compensation due under this Convention from any or all of the launching States which are jointly and severally liable". Hence, it is *first* reiterated that the Greek State has a right of recourse against any co-launching State, as clearly provided for in international space law.

However, law n° 4508/2017 also clarifies that the right of the Greek State to exercise its rights under Art. V para. 2 of the Liability Convention—laying down that a launching State may appeal to the court against any other co-launching State (s)—does not prevent it from using its right of recourse established in Art.11.[154] Importantly, it is at the same time underlined that said provision (namely, Art. V para. 2 of the Liability Convention) does not constitute a precondition for the Greek State to take legal action against any of the space operator(s) involved.

Eventually, the Greek space law stipulates that the State has a direct right of action against the insurer of the space operator(s), up to the amounts established in accordance with Art. 11.3 of law n° 4508/2017.[155] In such cases, the Greek State may be designated as a joint beneficiary to directly receive the insurance payment or any other guarantee. For this purpose, it is clearly mentioned that the space operator is required to provide—on a regular, i.e. annual basis—a copy of the insurance notice, or an attestation from the insurance company that said insurance is in force [1: Art. 11.6].

Having enacted measures to regulate the set up and implementation of space mission, as well as to address any risks and challenges that may arise in the context of such space operations (i.e., conducted under the jurisdiction of the Greek State), the legislator finally adopted rules to ensure the successful completion of space activities. As a result, provision was made to allow for the tracking of space

[154] This clause mainly concerns the rules established in Art.11.2 and Art. 11.3 of the Greek space law.

[155] In the same vein, "the Netherlands can exercise its right of recourse directly to the insurer up to the sum insured" [2: 73].

objects—throughout their entire operational duration within the context of space activities—, and to regulate of any other specificities related to the return of space objects, bearing in mind the need to avoid complications that could possibly occur even at this last stage of space activity.

5.4 Tracking, Return and Re-entry of Objects: Completion of the Activity

As already noted, the agreement reached—following the lengthy OST negotiations—was that space activities may be carried out either by governmental agencies or by non-governmental entities,[156] provided they are conducted under the continuous supervision of the States. In truth, despite being addressed by the same article (Art. VI of the OST), these two types of space activities remain fundamentally different, especially in reference to the depth and extent of State controls exercised over them.[157]

From a practical perspective, the obligation to exercise continuous supervision on both types of space activities raised concerns about its general implementation, especially in cases where space activities are being carried out by private operators.[158] Indeed, the OST provides no real "guidance on the scope, development, or implementation of supervision standards" while, to make things worse, "the transmission of commands to the spacecraft may be difficult to detect or decipher [*in addition to the fact that*] direct observation of all operations on orbit is not feasible" [46: 79]. More crucially, this issue has been compounded by the increased involvement of—public and private—operators and the growing utilization of space resources, especially through the extended use of new technologies like small satellites (which seem to pose challenges even as regards their registration [23]).

[156] "During these negotiations [*i.e., negotiations of the OST*], the Soviet's initial position was to limit space activity to government agencies and the United States' position was to open space to private entities. The negotiations resulted in both government agencies and non-governmental entities active in space, but under the expressed authorization and continued supervision of the state" [46: 77].

[157] And although States bear international responsibility for both types of space operations, in line with the OST, Art. VI.

[158] In fact, it was rightly noted that: "States appropriate funds for [*their*] space programs, albeit obscured [*by*] national defense and domestic security interests. (..) government officials, or their contracted representatives, participate in the planning, construction, operation, and supervision of their governmental space programs to ensure a successful outcome. Conversely, commercial entities operate out of self-interest; (. . .) they are subject to the general supervision exercised by government to ensure compliance with its common regulations, such as, labor standards, environmental protection, revenue collection (. . .) These regulations secure domestic interests, not the type of interests the foreign state signatories to a public international treaty seek to protect. Therefore, the OST specified the twin requirements of authorization and continuing supervision for commercial space activities" [46: 78].

Therefore, in this context, beyond the specific obligation to take measures to avoid any risks to life, the environment and assets presented above,[159] a duty to register space objects was adopted. This obligation was intended to essentially increase awareness and control over space objects, as well as to improve the safety of all space operations *per se* (i.e., considered independently from the idea that a threat could materialize in space). Thus, it is noted that said obligation was rooted on the perception that information-sharing—regarding space objects and/or orbital events—is vital for the conduct of *all* space activities.[160]

More precisely, a general registration requirement for all space objects was adopted first in the early 60s [47], without however specifying (then) the nature and scope of the information that had to be provided [23]. The duty to register space objects was implicitly though clearly reaffirmed in Art. V[161] and Art. VIII[162] of the OST, and later regulated more comprehensively in the *Registration Convention* signed in 1975 [48]. Adherence to this practice increased over time—even among States that were not, initially, parties to the Registration Convention[163]—and remains an overarching concern of the international community, to practically ensure as much control as possible over space objects (both at the level of States and at the international level).[164]

Finally, the return of objects launched in space was regulated with the adoption of the *Rescue Agreement,* in 1968 [49], applying in conjunction with all relevant provisions established in the other international space law treaties. Nonetheless, the observation that "(t)he scope of these treaties is far from comprehensive; it would be impossible to envision and provide for all the variables encountered in space exploration" [50] applies equally to the Rescue Agreement. As a result, many national legislators—as in the case of Greece—have also enriched the content of this last set of rules.

---

[159] See Sect. 5.3.

[160] The general idea of informing the competent authority of the UN or other actors involved in space activity is also mentioned in other articles of the OST, such as Art. XI: "In order to promote international cooperation in the peaceful exploration and use of outer space, States Parties (...) agree to inform the Secretary-General of the United Nations as well as the public and the international scientific community, to the greatest extent feasible and practicable, of the nature, conduct, locations and results of such activities.". For instance: "(t)he orbital parameters of the satellite (...) give some indication of the suitability of the satellite for various missions" [23: 411].

[161] Namely, States must safely and promptly return astronauts—in the event of an accident, distress, or emergency landing—*to the State of registry* of their space vehicle (see OST, Art. V).

[162] The State of registry of a space object "shall retain jurisdiction and control over such object, and over any personnel thereof, while in outer space or on a celestial body", see OST, Art. VIII.

[163] By way of illustration, it is mentioned that Luxembourg registered its space objects before acceding to the Registration Convention in 2021 [23: 407].

[164] Importantly, it is also argued that "States have been doing so believing that it is their legal obligation to comply with the Resolution (opinio juris)" and that the rule evolved to become a part of customary international law [23: 407].

5.4.1 Rules Applying to the Procedure for Registering Space Objects

The particular concept of a space object is defined—in Art. I.(b) of the Registration Convention—in a broad manner.[165] More precisely, the provision refers to "component parts of a space object as well as its launch vehicle and parts thereof". Hence, it may generally be regarded to include "any tangible human-made material or physical object or device, irrespective of its size, shape, composition and purpose (e.g., a payload or satellite, a launch vehicle or rocket, an astronaut suit, oxygen tank and other life support equipment, etc.)" [23],[166] which also leaves some margin of discretion to the national legislators.

Be that as it may, according to international space law, States are under the express obligation to register space objects, as long as they do qualify as launching States. Put differently, the *State of Registry* is defined, in Art. I.(c) of the Registration Convention, as any "*launching State* on whose registry a space object is carried in accordance with article II".[167] Therefore, a space activity conducted under the jurisdiction of the Greek State must *first* be in line with the international space law framework on the registration of space objects, as laid down in the Registration Convention or in any other related instrument (such as the 2007 Resolution on Enhancing Registration [22]) which may provide guidance on this matter; but also (*second*) with any additional rules that were adopted by the Greek legislator to further strengthen this specific framework.

5.4.1.1 International Space Law Framework: Rules on the Registration Procedure

According to the Registration Convention, if there is only one launching State, that State is bound to "register the space object by means of an entry in an appropriate registry which it shall maintain[168]"; in cases where there are two or more launching States, they must jointly determine which one of them shall register the space object.[169] Art. II para. 3 underlines that the technical modalities of such registration process, like the contents of the registry and the conditions under which it is

[165] This has led to different interpretations of the term: *inter alia* "Some States (such as the US and France) have interpreted "space object" to include non-functional objects, such as discarded rocket stages and debris, while others (e.g. Russia) consider only payloads" [23: 407].

[166] "In this context, Art. II provides further clarifications, by stipulating that any object "launched into Earth orbit or beyond" ought to be registered, which could as well be interpreted as meaning that "(i)f a space object is not launched 'into earth orbit or beyond' it would not be required to be registered; e.g., an object sent only on a sub-orbital flight" [23: 407].

[167] Emphasis added.

[168] Registration Convention, Art. II.(1).

[169] As further established in the Registration Convention, Art. II.(2).

maintained, are to be established by each State of registry concerned. At the same time, and no less importantly, it is stipulated that the State of registry must also provide all related information to the Secretary-General of the United Nations, as soon as practicable.[170]

As a result, it follows from these articles of the Registration Convention that States parties have two key obligations. More precisely, States must—at the national level—(i) establish and maintain a national registry for their space objects, which may be further divided into (i.a) an obligation to establish and maintain an appropriate national register for space objects and (i.b) an obligation to record in this register space objects [23]; and (ii)—at the international level—, the obligation to inform the Secretary-General of the United Nations about the establishment of a national registry, as well as to notify said authority of all related information about the space objects listed in their national registry (including any changes to the information initially provided).

In turn, the Secretary-General of the United Nations shall maintain its own register, containing all the information that States parties are bound to provide regarding their space objects launched into Earth orbit or beyond. Interestingly, the Registration Convention (in Art. III and/or Art. IV) does not grant "any discretionary authority of the UN to examine the accuracy or question the inaccuracy of the information supplied by the State Parties to the Registration Convention" [23: 412]. Therefore, the Secretary-General of the United Nations has no control over the information provided, or any obligation of any kind to conduct such control.

In this context, it is noteworthy that the first obligation of States parties (namely, (i) the obligation to establish and maintain a *national* registry for space objects, mentioned in Art. II para. 1 and 3 of the Registration Convention) allows sufficient discretion for States to adapt it to their particular circumstances, taking into account international practices in this field. On the contrary, the second one—that is, (ii) the obligation to inform the Secretary-General of the United Nations—is regulated in detail. Indeed, Art. IV para. (1) of the Registration Convention enumerates the pieces of information (for each space object) that must be notified to the United Nations authority: the article precisely mentions (a) the name of launching State or States; (b) an appropriate designator of the space object or its registration number; (c) the date and territory or location of launch; (d) the basic orbital parameters, including: (i) the nodal period; (ii) the inclination; (iii) the apogee; (iv) the perigee and (e) the general function of the space object.

In addition to the above, a general obligation (upon States of registry) to provide follow-up information—i.e., each time such information becomes available and is necessary—was laid down (regarding space objects carried on their national registries, as well as space objects concerning which information had been previously duly transmitted to the United Nations authority but are no longer used in Earth

[170] Registration Convention, Art. IV.(1).

orbit[171]). Eventually, a few additional and more specific rules were also established to provide solutions in cases where a space object that had caused damage to a State or to any of its natural or juridical persons (or which may be of a hazardous or deleterious nature) is not identifiable.[172]

5.4.1.2 Soft Law Instruments and Guidelines for Registering Space Objects

On a more practical level, one direct result of the above is that "(p)rivate companies are not entitled to directly register their space objects with the UN Secretary-General. In order to do so, they have to rely upon their respective State(s)" [23: 408]. Conversely, it remains the responsibility of States to ensure that private space operators will indeed provide them all the necessary information required for inclusion in the national register, as well as for communication to the UN Secretary-General. Hence, considering the deep complexity (and rapid pace of change) of practice patterns as well as of space activities in general,[173] soft-law instruments were adopted to provide additional guidance.

In particular, the UN General Assembly adopted—in 2007—Resolution 62/101, to *initially* encourage States to sign, ratify and implement the Registration Convention, in order to contribute to "common procedures, at the national and international levels, for registering space objects with the Register", as well as to promote "uniformity with regard to the information to be furnished and recorded in the Register concerning space objects listed in the appropriate registries" [22: 2]. Thus, the resolution provided specific guidance on the information to be supplied for the registration of space objects with the UN Secretary-General. More precisely, Art. 2(a) of the 2007 Resolution on Enhancing Registration clarified the information that could be provided on the basis of Art. IV (1) of the Registration Convention. For example, explicit suggestions were made to communicate (i) the COSPAR ID (i.e., the Committee on Space Research international designator) of the space object,

---

[171] Registration Convention, Art. IV (2): "Each State of registry may, from time to time, provide the Secretary-General of the United Nations with additional information concerning a space object carried on its registry" and Art. IV.(3): "Each State of registry shall notify the Secretary-General of the United Nations, to the greatest extent feasible and as soon as practicable, of space objects concerning which it has previously transmitted information, and which have been but no longer are in Earth orbit".

[172] See, Registration Convention, Art. VI.

[173] For instance, "changes in space activities since the Registration Convention entered into force include the continuous development of new technologies, an increase in the number of States carrying out space activities, an increase in international cooperation in the peaceful uses of outer space and an increase in activities carried out by non-governmental entities, as well as partnerships formed by non-governmental entities from more than one country", [22: 2].

where appropriate;[174] similarly, it was recommended to use the "(c)oordinated Universal Time as the time reference for the date of launch[175]" as well as "(k)-ilometres, minutes and degrees as the standard units for basic orbital parameters[176]".

Finally, the 2007 Resolution on Enhancing Registration contained suggestions as to the information that should be notified to the UN Secretary-General under Art. IV para. (2) of the Registration Convention.[177] By way of illustration, mention was made to the geostationary orbit location, again where appropriate [22: Art. 2 (b) (i)]; any change(s) of status in operations (*inter alia*, when a space object is no longer functional) [22: Art. 2 (b) (ii)]; the approximate date of decay or re-entry, if States are able to verify that information [22: Art. 2 (b) (iii)]; the date and physical conditions of moving a space object to a disposal orbit [22: Art. 2 (b) (iv)] and, lastly, Web links to official information on the space objects concerned [22: Art. 2 (b) (v)].

Against this background, the Sofia Guidelines proposed specific draft rules to assist States in setting up their national frameworks. Hence, draft Art. 10 titled *Registration* provided for the creation of a national registry for space objects,[178] and enumerated the pieces of information that could/should be included in such an instrument. More precisely, said provision makes reference to (a) the name of the launching State or States (including the name of any private launching entity, whether a natural or legal person); (b) the registration number of the space object; (c) the date and territory or location of the launch; (d) the basic orbital parameters including nodal period, inclination, apogee and perigee; (e) the general function of the space object as well as (f) any other additional information, in line with international space law.[179]

At the same time, the obligation to update all relevant information was reiterated in draft Art. 10.7, stipulating that: "*(a)ny relevant change* with regard to the information mentioned in paragraph 1 should be registered in the National Registry. The Secretary-General of the United Nations shall be informed accordingly[180]".

---

[174] This is a clarification of Art. IV (1) of the Registration Convention, mentioning that "Each State shall furnish ... (b). *An appropriate designator* of the space object or its registration number", (emphasis added).

[175] This provision clarified Art. IV (1) of the Registration Convention, mentioning that "Each State shall furnish ... (c). Date and territory or location of launch".

[176] Art. IV (1) of the Registration Convention lays down that States must provide information about "(d). Basic orbital parameters", in general.

[177] As mentioned above, the Registration Convention, Art. IV (2) lays down that States "may, from time to time, provide the Secretary-General of the United Nations *with additional information* concerning a space object carried on its registry", (emphasis added).

[178] Sofia Guidelines, draft Art. 10.1: "A National Registry is hereby established for the registration of space objects. The authority (namely the competent minister, preferably the same as in article 3) shall maintain the national space register".

[179] Sofia Guidelines, draft Art. 10.5: "Additional information and information *in accordance with* the Registration Convention and/or the United Nations Registration Practice Resolution as specified in an implementing decree/regulation shall also be included in the National Registry", (emphasis added).

[180] Emphasis added.

Overall, draft Art. 10 essentially reproduced the wording of Art. IV of the Registration Convention and built upon the clarifications provided by the 2007 Resolution on Enhancing Registration. Importantly though, a comment under this draft provision underlined that it remains the responsibility of States to get the respective information from all operators, making it clear that it is up to the national legislator to actually establish efficient measures in this respect.

Therefore, the Greek legislator adopted specific rules to meet the two obligations incumbent on the Greek State, on the basis of the Registration Convention: namely, to establish and maintain a national registry for space objects under its jurisdiction; and to inform the UN Secretary-General about its establishment as well as the space objects listed therein.

5.4.1.3 The Obligation of Greece to Establish and Maintain a National Register

In line with Art. 17 of the Greek space law, a National Register for Space Objects [*Εθνικό Μητρώο Διαστημικών Αντικειμένων*] was created, to be maintained and updated by the ministry that was competent for space matters at that time, namely the Ministry of Digital Policy, Telecommunications and Information [*Υπουργείο Ψηφιακής Πολιτικής, Τηλεπικοινωνιών κι Ενημέρωσης*]; the competent authority being, today, the Ministry of Digital Governance. It was also established that said register will contain information on space objects which are used in the context of authorized space activities [1: Art. 17.1], for which Greece is regarded as a launching State (either on its own, or together with another State, or with other States). Importantly, it was also clarified that the obligation to register the space object—when Greece is a co-launching State—does not apply in case the object has already been, or will be, registered by another State or International Organization, in accordance with the Registration Convention.

On a practical level, the registration of a space object is performed following an examination of the relevance and validity of all the information provided in relation to the object, which must be submitted by the operator. Following the completion of the registration procedure, the space object will be listed in the National Register for Space Objects (where all related information is noted) and a unique national registration and listing number [*Εθνικός Αριθμός Καταχώρισης και Καταγραφής*] will be assigned to each object, through a decision by the competent authority, which is today the Minister of Digital Governance [1: Art. 17.2 (α)]; as was recently done, for instance, in the case of the registration of the satellite *Hellas Sat 4* [51].

More precisely, according to Art. 17.2 (β), all those pieces of information referred to in Art. IV of the Registration Convention must be notified to the competent authority by the operator, in addition though to those enumerated in the *Model Registration Form for Registration of Space Objects with the Secretary-*

General [52].[181] Quite interestingly, the suggestion was made—i.e., during the public consultation on the draft of the Greek space law—to additionally list, in the National Register, *inter alia* companies whose products and/or services are provided by satellites, or which provide other space-related services; the aim being to achieve a full understanding of all the space and space-related activities conducted under the supervision of the Greek State, and to also enable the adoption of space policies that would account for all the interests at stake [13: Art. 1]). This proposal was not retained in the final text of the provision concerned.

Nonetheless, law n° 4508/2017 clarified that the operator shall also communicate to the Greek competent authority information relating to: (α) the manufacturer of the space object; (β) the company launching the space object; (γ) the operator and the owner of the space object; (δ) the major component elements and parts inside the space object; (ε) the ITU filings under which the object will operate, and (στ) the authorization (to engage in space activity) which has been granted to the space operator together with any other related supporting documentation [1: Art. 17.3].

Most importantly, the space operator must notify all the information referred to in the relevant article of the law to the competent space authority (today, the Minister of Digital Governance), without undue delay.[182] In case of any change regarding the information that was provided in the initial application—i.e., possibly affecting the conditions under which the authorization was granted—, a correction must be made by means of an additional inscription; and the inscription of the new information must be requested by the space operator within a delay of thirty (30) days, starting from the date on which they became aware of the change(s) in question. Hence, in the event that the space operator does not communicate such modifications within this precise period of time, the competent authority will be empowered to revoke or suspend the authorization which had been granted [1: Art. 17.4].

Finally, in line with Art. 17.5 of law n° 4508/2017, the registration of space objects shall be made after their launch and following a written notification sent to the competent authority (i.e., to the Minister of Digital Governance) by the operator. Such a document must precisely indicate the launch time, the launch activity and the placing into operation of the space object(s). Importantly, the law clearly mentions that the registration may only concern space activities that have been authorized [1: Art. 17.1]; therefore, such registration is the final step in completing the launch of a space object, within the context of a space activity duly authorized by the Greek competent authority. Furthermore, to ensure a smooth transition to the new regime

[181] Law n° 4508/2017, Art. 17.2 (β) mentions: "The information to be recorded shall include those pieces referred to in Article IV of the Convention on Registration of Objects Launched into Outer Space, *as well as those* mentioned on the Model Registration Form for registration to the UN Register", (emphasis added).

[182] Although it was not explicitly mentioned in the Greek space law, it is logical to consider that this reasonable period starts at the time when the authorization to conduct the space activity is granted (taking into account that Art. 17.1 of the Greek space law makes reference to the obligation to register the space objects related to *authorized* space activities).

regulating space activities, Art. 17.6 of the Greek space law established that—following the adoption of a decision by the competent authority for these matters (thus, for decisions taken at that point, by the then competent minister i.e., the Minister of Digital Policy, Telecommunications and Information [*Υπουργός Ψηφιακής Πολιτικής, Τηλεπικοινωνιών κι Ενημέρωσης*])—, space objects relating and/or involved in any space activities which were in progress at the time of entry into force of law n° 4508/2017, would likewise be registered into the National Registry for Space Objects.

5.4.1.4 Duty of the Greek State to Inform the United Nations Secretary-General

The second informational obligation placed upon States parties by Art. II of the Registration Convention is, *first of all*, to inform the Secretary-General of the United Nations of the establishment of such a register (i.e., of a national register). In addition to that, States are also required to furnish said authority with information on space objects carried on their national register, as laid down in Art. IV of the Registration Convention. The rationale for creating these duties was to enhance security and transparency through collaboration between all States, while also achieving standardization of the process for gathering information. In fact, as the latter goal appeared to be critically important, the UN OOSA also provided States with a *Model Registration Form* to assist them—as well as any organization(s) concerned—in registering their space objects (as was requested in the 2007 Resolution on Enhancing Registration), which is explicitly referred to in the Greek space law [52; 1: Art. 17.2.(β)].

Hence, the Greek State became subject to this obligation after it implemented its initial space activities, and thus became a space-faring country—which took place, as mentioned in the 2021 official statement submitted by Greece to the Scientific and Technical Subcommittee of the UN COPUOS (UN COPUOS STSC) [53]—in 2003, following the launch its first Telecommunication Satellite *Hellas Sat 2*. Subsequently, in June 2017, a second satellite was launched—under the name *Hellas Sat 3*—to replace the aging *Hellas Sat 2*; and in February 2019, *Hellas Sat 4* followed.

As a result, to bring its practice into line with its international obligations, the Greek administration informed the Secretary-General of the United Nations by note verbale—in 2004 and 2008—that it "maintains an appropriate registry of objects launched into outer space" (a precise reference was made, in the document, to Art. II of the Registration Convention) [54, 56]. Therefore, the United Nations Secretary-General was informed about the maintenance of national records and objects launched into space since 2004, that is prior to the detailed regulation of such a registry for space objects launched under the jurisdiction of the Greek State (which was introduced later, in the Greek space law adopted in 2017).

In truth, the practice of States in the field of space activity may explain this discrepancy, *as on the one hand*, States which have ratified the Registration

Convention (as in the case of Greece, which ratified said convention in 2003 [55]) are under the duty to comply with the information obligations laid down therein; *on the other hand*, it was clearly recommended—in the 2007 Resolution on Enhancing Registration—that "States that have not yet ratified or acceded to the Registration Convention should become parties to it in accordance with their domestic law *and, until they become parties, furnish information* in accordance with General Assembly resolution 1721 B (XVI)", [22: 2, para. 1(a)].[183] On this basis, and in more practical terms, "(e)ven States which have not established national registries under Article II (1) have often nevertheless sent registration information on their satellites to the UN[184]"; the Greek administration adhered to this practice as well.

More precisely, Greece has up until now submitted to the Secretary-General of the United Nations information on three (3) space objects operating under its jurisdiction. The first notification—sent in the early 2000s and referring to Art. IV of the Registration Convention—, was the note verbale dated 29 March 2004. The document contained information concerning the launch of the satellite named *Hellas Sat 2*, which took place on 13 May 2003 from Cape Canaveral in Florida, United States; in this regard, the note verbale precisely mentioned all the information explicitly enumerated in Art. IV para. 1 of the Registration Convention [56].

Subsequently, the Greek State submitted to the UN Secretary-General a second note verbale—dated 2 March 2022 [57]—to provide information on the launch of two other satellites, *Hellas Sat 3* and *Hellas Sat 4*. The information transmitted related to the fact that the satellite *Hellas Sat 3* had been launched under the COSPAR designator 2017–040A and the national registration number 1002. Greece was indentified as the State of Registry (and thus as a launching State as well), while Cyprus, France, the United Kingdom of Great Britain and Northern Ireland were referred to as co-launching States. The above note verbale also contained the information referred to in the Registration Convention, Art. IV para. 1 and para. 2, such as *inter alia* the precise date and territory or location of the launch (i.e., 28 June 2017 at 2115 h 0 s UTC; Guiana Space Centre, French Guiana) and additional voluntary information to be mentioned in the Register of Objects Launched into Outer Space (that is, the Geostationary position; the Space object owner or the space operator; the operator's Website and the Launch vehicle).

In the same note verbale dated 2 March 2022, the Greek State provided information also about *Hellas Sat 4*. The satellite had been launched under the COSPAR designator 2019–007A and the national designator/registration number 1003; Greece was referred to as the State of Registry (and launching State), and Cyprus and Saudi Arabia were mentioned as co-launching States. Once again, the note verbale additionally contained the information enumerated in Art. IV para. 1 and para. 2 of the Registration Convention, such as the date and territory or location of the launch (5 February 2019 at 2101 h 0 s UTC; Guiana Space Centre, French

---

[183] Emphasis added.

[184] "It is estimated that about half of the notifying States sent their first registration information 4 or more years prior to the notification of the registry under Article II(1)" [23: 409].

Guiana); the basic orbital parameters; the general function of the space object and other voluntary information to be included in the Register of Objects Launched into Outer Space [57].

More importantly, as the aim of the obligation to provide information on space objects being launched was to improve the safety and security of space activities—i.e., in a concrete and tangible manner—, Art. III para. 2 of the Registration Convention additionally states that "(t)here shall be full and open access to the information in this Register", namely the register for space objects held by the Secretary-General of the United Nations. On a more practical level, "(f)or many years access to a UN repository library was the only avenue to inspect the data (...). Recently though, the UNOOSA has created a web-based Online Index of Objects Launched into Outer Space, which contains information obtained from non-official sources, particularly about those space objects that have not been notified by States (...). In October 1994, an online tabulation and analysis of the registration documents were made public" [23: 413].

As a result, the launching and operation of satellites—under the jurisdiction of the Greek State—which has been notified to the competent UN authority is also entered on the *Online Index of Objects Launched into Outer Space* [58]. This indeed appears to be a quick and efficient means of accessing all the information provided to the United Nations by States, in line with Art. III of the Registration Convention and the General Assembly resolution 1721 B (XVI), as is shown in the Fig. 5.1 below.

5.4.2 The Return of Astronauts and the Re-entry of Space Objects

Finally, during the discussions that led to the adoption of the OST, special focus was placed on potential facets of space activity which must however be addressed, bearing in mind that the ultimate purpose is to efficiently ensure *the full control of all aspects* (i.e., towards its final and safe completion). Therefore, two additional topics were regulated in the OST, and subsequently in more detail by the Rescue Agreement [49]: (i) the necessity to provide assistance to, and ensure the return of, astronauts; and (ii) the need to recover space objects and their components following their re-entry into the Earth's atmosphere, since they may sometimes return to Earth—or splash into the oceans—in an uncontrolled manner (which, in truth, is far from being an unusual phenomenon in the context of space activity [50, 59]).

More precisely, it was established that, *on the one hand,* States parties must provide any possible assistance to astronauts "in the event of an accident, distress or emergency landing", as they are the envoys of mankind in space,[185] such as by

[185] OST, Art. V: "States Parties to the Treaty *shall regard astronauts as envoys of mankind in outer space* and shall render to them all possible assistance in the event of accident, distress, or emergency landing on the territory of another State Party or on the high seas ...", (emphasis added).

Online Index of Objects Launched into Outer Space

▶ FILTER BY ...

Important Note: Information in square brackets ([and]) and highlighted in green has been obtained from other sources and has not been communicated officially to the United Nations. Reference to external websites does not imply endorsement by the United Nations Office for Outer Space Affairs (UNOOSA) of their contents. The views expressed are those of the authors and do not necessarily reflect the policies or views of UNOOSA. The hyperlinks are provided solely for informational purposes.

Search Object

additional criteria:

State / Organizaton: *Greece* × State / Organizaton: *Greece* ×

found 5 Objects

Clear All Criteria

International Designator	National Designator	Name of Space Object	State/ Organization	Date of Launch	GSO Location	UN Registered	Registration Document	Other Documents	Status	Date of Decay or Change	Function of Space Object	Secretariat`s Remarks	External website
[2022-122*]		[QUBIK 4]	[(for Greece)]	[2022-10-01]		No			[decayed]	[2022-10-05]	------	Not registered with the United Nations. International Designator pending assignment.	
[2022-122*]		[QUBIK 3]	[(for Greece)]	[2022-10-01]		No			[decayed]	[2022-10-05]	------	Not registered with the United Nations. International Designator pending assignment.	
2019-007A	1003	HS4SGS1 (HELLAS SAT-4) [(SAUDIGEOSAT 1, SGS 1)]	Greece	2019-02-05	+039 degrees East	Yes	ST/SG/SER.E/1044		in GSO		Telecommunication geostationary satellite that operates in Ku-band and Ka-band	------	Link
2017-040A	1002	HS3IS (HELLAS SAT-3) [INMARSAT S EAN]	Greece	2017-06-28	+039 degrees East	Yes	ST/SG/SER.E/1044		in GSO		Telecommunication geostationary satellite that operates in Ku-band, Ka-band and S-band	------	Link
[2003-020A]	------	HELLAS-SAT 2	Greece	2003-05-13	[+039 degrees East]	Yes	ST/SG/SER.E/446		in GSO		Geostationary telecommunication satellite at 39 degrees East longitude	------	Link

Fig. 5.1 UNOOSA/Online Index of Objects Launched into Outer Space, State/Organization: Greece—accessed July 2024. Note that QUBIK-3 and QUBIK-4 are two Greek pico-satellites, measuring 5 by 5 by 5 centimeters, manufactured in Greece and put into space by the non-profit organization Libre Space Foundation and in orbit by a Firefly Aerospace's Alpha Flight 2 rocket

ensuring their safe and prompt return to the State of registry of their space vehicle.[186] *On the other hand*, it was also laid down that any space objects and/or their components found anywhere on Earth should be returned to their State of Registry (OST, Art. VIII[187]). These two topics were further regulated in the Rescue Agreement, which is regarded as having essentially a humanitarian purpose, based on the fact that primary importance is given there to the safe return of astronauts [50: 474]. As a matter of fact, it is worth noting that more articles address the rescue of astronauts (precisely, Art. 1 to 4 of the Rescue Agreement; "in other words the bulk of its substantive operative provisions" [60: 424]) compared to the return of space objects, which is mainly regulated by Art. 5. In addition to that, the key topic of compensation for possible damage caused by a space object (either in space or on

[186] See OST, Art. V para. 1.

[187] OST, Art. VIII: "Such objects or component parts found beyond the limits of the State Party to the Treaty on whose registry they are carried shall be returned to that State Party, which shall, upon request, furnish identifying data prior to their return."

Earth) is thoroughly addressed in another treaty precisely dedicated to this subject, namely the Liability Convention.

However, a closer examination reveals that, in reality, the Rescue Agreement seems to have established a less generous system of protection for astronauts, than the one initially laid down in the OST. In this vein, it is underlined that "the reference to assistance by astronauts of one state to astronauts of other states in outer space under the second paragraph of Article V of the Outer Space Treaty does not reappear in the Rescue Agreement. Articles 1 through 4 of the Rescue Agreement exclusively concern terrestrial events" [60: 422–423]. Parallel to that, developments in the field of space have shown that States practically paid more attention to the recovery and return of space objects.

Consistent with this general trend, the Greek space law did not include any specific additional provisions on the rescue of astronauts—the rules established in the Rescue Agreement are, in any case, binding on all States parties, including Greece—, but rather focused on the issue of the recovery and return of space objects [1: Art. 13]. Hence, the following analysis will likewise place emphasis on the latter topic.

5.4.2.1 International Law Applying to the Recovery of Space Objects

As regards the recovery of space objects, in case of their return to Earth in a territory under another State's jurisdiction—or on the high seas or in any other place not under the jurisdiction of any state—, the rules adopted in the OST and the Rescue Agreement are of a general nature, still they need to be implemented by national authorities. Hence, given their duty to exercise continuous supervision over their space activities (including over their space objects),[188] States must regularly notify the UN Secretary General of any information concerning space objects carried on their registry, as established in Art. IV para. 2 and 3 of the Registration Convention (the purpose being to allow for a rapid and efficient response in case damage finally occurs, either in outer space or after the re-entry of space objects in the atmosphere).

In this context, Art. 5 para. 1 of the Rescue Agreement clarified that any State party which receives information or discovers that a space object—or any of its component parts—has landed or been found in areas under its jurisdiction, shall *first* inform (i) the launching authority, and then (ii) the Secretary-General of the United Nations. Each State party is also committed to act likewise in case it receives information that a space object or any of its component parts is found on the high seas or in any other place not under the jurisdiction of any State.

Furthermore, in the event that a State party discovers a space object in an area under its jurisdiction, it must also take such steps *as it finds practicable* to recover the object or its component parts. However, the measures or steps to be taken are not listed in the Rescue Agreement or in any other binding international space law

---

[188] See OST, Art. VI, read in conjunction with OST, Art. VII and Art. VIII.

instrument; thus, the decision is left to the discretion of the States that are each time concerned. Art. 5 para. 2 underlines that the State party shall do so upon the request of the launching authority, with its assistance. In this regard, the launching authority may be requested to furnish identifying data prior to such return,[189] as initially established in Art. VIII of the OST.

Importantly, the Rescue Agreement recognizes the possibility that a space object, or its component parts may be of a hazardous or deleterious nature, and lays down an affirmative duty for the launching authority to act immediately [50: 475]. Indeed, Art. 5 para. 4 regulates the discovery and return of such elements, by way of derogation from paragraphs 2 and 3 of Art. 5; said provision stipulates that if a State party has reasons to believe that a space object found in an area under its jurisdiction (or recovered by it elsewhere), is toxic or hazardous, it may so notify the launching authority. In turn, the launching authority shall *immediately take effective steps*—under the direction and control of the State party which found the space object or its components—to eliminate any possible danger of harm.[190]

As the Rescue Agreement requires that States parties return foreign space objects discovered on their territories to their owners, but also notify the Secretary-General of the United Nations about discovered objects, the UNOOSA maintains a list of recovery notifications which is publicly available[191] [62], to essentially encourage transparency and collaboration between States in the conduct of space activity. According to this database, no notification from the Greek administration has up to now been communicated to the United Nations authority.

5.4.2.2 Clarifications that Were Introduced in Art. 13 of the Greek Space Law

To strengthen the international space law framework, the Greek space legislation laid down a few clarifications in its Art. 13.[192] As already mentioned, the emphasis was placed on regulating the recovery and return of space objects—rather than elaborating on the issue of the rescue of astronauts—, and said provision of law n° 4508/2017 entitled *Return of space object* does not contain any rules on the safe return of astronauts.

In essence, Art. 13 para 1 stipulates that a space object intercepted or found on Greek territory—or any other area under Greek jurisdiction—, must be returned without any delay to the competent authorities, which shall immediately notify the

[189] The Rescue Agreement, in Art. 5 para. 3, clarifies that "... objects launched into outer space or their component parts found beyond the territorial limits of the launching *authority shall be returned to or held at the disposal of representatives of the launching authority*, which shall, upon request, furnish identifying data prior to their return", (emphasis added).

[190] As a rule, the costs incurred in satisfying the obligation to recover and return a space object shall be borne by the launching authority, see Rescue Agreement, Art. 5 para. 5.

[191] United Nations Office for Outer Space Affairs [61].

[192] Interestingly, the Sofia Guidelines did not provide any further guidance on the matter.

competent minister (today, Minister of Digital Governance). In turn, the competent minister shall arrange for the return of said object to the State of registry, in line with what is stipulated in the Registration Convention. Most importantly, it was established that the provision applies without prejudice to any measures adopted to ensure the safety and protection of assets and persons [1: Art. 13.1]. However, to leave the competent authorities with a certain margin of discretion, the article remained silent as regards the specific form and the content of such measures.

Should additional investigation be required to determine who the State of Registry or the Launching State is—such as when more than one State is involved—, Art. 13 para. 2 lays down that the competent authority is entitled to take interim measures, at its discretion. The law does not clarify what kind of measures could be taken, as general reference is made to "*any measure necessary for the preservation* of the object or the objects found or identified[193]". In the same vein, the Greek space law stipulates that in those instances where the space object is returned to the competent authority—or, if required, *before* such return—, any necessary measures shall be taken for the protection of victims' rights, as well as for the protection of property which was damaged by the space object" [1: Art. 13 para. 3][194]". Importantly, the Explanatory Memorandum to the Draft of law n° 4508/2017 additionally mentioned the possibility of collaboration between (precisely) the competent authority, the Civil Protection and National Defense services, to eventually obtain information concerning the owner and the operator of the space object [3: 4, comment under Art. 13].

Regarding the adoption of the rules presented above by the Greek legislator, one may argue that comments on similar provisions that were established in other domestic laws could be taken into account. In particular, it has been mentioned that "the obligations contained within the Rescue Agreement do not necessarily prompt states to enact national space legislation, which is why the obligations under the Rescue Agreement have not been incorporated into national space legislation (...). Only Belgium has included a provision that deals with the obligations under the Rescue Agreement (...). Although the inclusion of this provision provides additional clarity, it seems superfluous as the return of the space object will definitely alert state authorities and already ensure that the space object is in the possession of the competent authorities. Therefore, the state should have no problem in adhering to its obligations under the Rescue Agreement. The simple fact that no similar provisions have been included by other countries speaks volume about the necessity for such a provision in national space legislation" [2: 183–186].

On the other hand, the importance attached to the issue of security as regards space activities and space objects (namely, until completion of the operations) is indisputable, especially given the constantly growing use of space, and the increase in both space operators and space objects. Therefore, experts have also pointed out that regulating the return of space objects is "not a concern merely for

[193] Emphasis added.

[194] Emphasis added.

academic speculation, but rather a dilemma of the utmost practical importance", [50: 481] that should be addressed taking into account the scale and complexity of the issue [63]. Seen from this angle, any clarification or even reiteration of the applicable international space law rules may be useful, as it may at least raise awareness of the related security issues.

This final Section presented the rules of law adopted in the Greek space legislation to regulate the last phases of a space activity and completes the presentation of the measures that were adopted by the Greek State to regulate activities being carried out under its jurisdiction.

References

1. *Law n° 4508/2017 on the Authorization of space activities - Registration in the National Registry of Space Objects - Establishment of the Hellenic Space Agency and other provisions.* Published at the Greek Government Gazette A' 200/22.12.2017, p. 4411. et seq.
2. Froehlich, A., & Seffinga, S. (Eds.). (2018). *National space legislation – A comparative and evaluative analysis*. Springer.
3. *Explanatory Memorandum to the Draft law entitled Authorization of space activities - Registration in the National Registry of Space Objects - Establishment of the Hellenic Space Agency and other provisions"*. Accessed from https://www.hellenicparliament.gr/Nomothetiko-Ergo/Anazitisi-Nomothetikou-Ergou?law_id=ca706be4-0ccc-4dbf-a8ce-a83b01095b01 (in Greek).
4. Igoumenidi, T. (2022, May). *The National Programme on small satellites moves on to the next phase*. Naftemporiki. Accessed July 2024, from https://www.naftemporiki.gr/business/1333502/stin-epomeni-fasi-to-ethniko-programma-mikron-doryforon/
5. Delevegos, D. (2023, February 13). *Country to launch mini-satellites the size of a washing machine*. Ekathimerini-com. Accessed July 2024, from https://www.ekathimerini.com/economy/1204549/country-to-launch-mini-satellites-the-size-of-a-washing-machine/
6. Taubenfeld, H. J. (1970). Outer space: The "territorial" limits of nations. *Fordham Law Review, 38*, 1–2.
7. UN COPUOS LSC. (2013, March 26). *Information on the activities of international intergovernmental and non-governmental organizations relating to space law. Note by the Secretariat.* UN Doc A/AC.105/C.2/2013/CRP.6. Accessed from https://www.unoosa.org/pdf/limited/c2/AC105_C2_2013_CRP06E.pdf
8. von der Dunk, F. G. (2012). *Towards 'flags of convenience' in space?* (p. 76). Space, Cyber, and Telecommunications Law Program Faculty Publications. Accessed from https://digitalcommons.unl.edu/cgi/viewcontent.cgi?article=1075&context=spacelaw
9. UNGA. UN COPUOS. (2021, February 19). *National legislation and practice relating to the definition and delimitation of outer space Note by the Secretariat – Addendum*. LSC 60th session. UN DOC A/AC.105/865/Add.25. Accessed from https://www.unoosa.org/oosa/oosadoc/data/documents/2021/aac.105/aac.105865add.25_0.html
10. UN COPUOS LSC. (2019). *Matters relating to the definition and delimitation of outer space: Replies of Greece. National legislation and practice relating to the definition and delimitation of outer space*. UN Doc A/AC.105/C.2/2019/CRP.6 (28 March 2019). Accessed from https://www.unoosa.org/res/oosadoc/data/documents/2019/aac_105c_22019crp/aac_105c_22019crp_6_0_html/AC105_C2_2019_CRP06E.pdf
11. UN COPUOS LSC. (2017). *Matters relating to the definition and delimitation of outer space: Replies of Greece*. UN Doc A/AC.105/C.2/2017/CRP.16 (28 March 2017). Accessed from https://www.unoosa.org/res/oosadoc/data/documents/2017/aac_105c_22017crp/aac_105c_22017crp_16_0_html/AC105_C2_2017_CRP16E.pdf

12. UNOOSA. *Treaty on principles governing the activities of states in the exploration and use of outer space, including the moon and other celestial bodies. GA Res 2222 (XXI), signed on 27 January 1967 (entered into force, 10 October 1967)*. Accessed from https://www.unoosa.org/oosa/oosadoc/data/resolutions/1966/general_assembly_21st_session/res_2222_xxi.html (*Outer Space Treaty; OST*).
13. Ministry of Digital Governance. *Public consultation procedure - Comments under draft law n° 4508/2017 (in Greek)*. Accessed July 2024, from http://www.opengov.gr/digitalandbrief/?p=771
14. *Law n° 4048/2012 (published at the Greek Government Gazette A' 34/23.02.2012, p. 915 et seq.) and amended by Art. 61 of law n° 4622/2019*. Published at the Greek Government Gazette A' 133/07.08.2019, p. 3223 et seq..
15. Kerkonian, A. D., & Raju, N. (2020). The legal challenges of international suborbital flights: A bilateral solution. *Journal of Air Law and Commerce, 85*(387) Accessed from https://scholar.smu.edu/jalc/vol85/iss3/2
16. GOV.UK. (2022, June 01). *Spaceflight legislation and guidance*. Accessed July 2024, from https://www.gov.uk/guidance/spaceflight-legislation-and-guidance#overview-of-the-outer-space-act-1986-and-the-space-industry-act-2018
17. UNCOPUOS. (2022, February 28). *Questions on suborbital flights for scientific missions and/or for human transportation Note by the Secretariat*. UN Doc A/AC.105/1039/Add.18. Accessed from https://www.unoosa.org/oosa/oosadoc/data/documents/2022/aac.105/aac.1051039add.18_0.html
18. SpaceNews. (2019, May 21). *Suppliers hope to tap into growing space market*. Jeff Foust. Accessed July 2024, from https://spacenews.com/suppliers-hope-to-tap-into-growing-space-market/
19. TechCrunchCom.Space. (2022, August 9). *Muon Space plans a 'turnkey solution' for custom Earth observation satellites*. Devin Coldewey. Accessed July 2024, from https://techcrunch.com/2022/08/09/muon-space-plans-a-turnkey-solution-for-custom-earth-observation-satellites/?guccounter=1
20. Enterprise Greece – Invest & Trade. (n.d.). *Greek companies participate at space tech expo 2022*. Accessed July 2024, from https://www.enterprisegreece.gov.gr/en/news/press-releases/space-tech-expo-2022
21. General Assembly Resolution. *59/115 Application of the concept of the 'launching state'*. UNGA 59th Session, UN Doc A/RES/59/115 (25 January 2005), (Resolution on the concept of 'launching state').
22. United Nations General Assembly. Resolution. *62/101 recommendations on enhancing the practice of states and International Intergovernmental Organizations in registering space objects*. UNGA 62nd Session, UN Doc A/RES/62/101 (10 January 2008), (Resolution on Enhancing Registration).
23. Jakhu, R. S., Jasani, B., & McDowell, J. C. (2018). Critical issues related to registration of space objects and transparency of space activities. *Acta Astronautica, 143*, 406–420.
24. von der Dunk, F.G. (1992). *Liability versus responsibility in space law: Misconception or misconstruction?* (p. 21). Space, Cyber, and Telecommunications Law Program Faculty Publications. Accessed from https://digitalcommons.unl.edu/spacelaw/21
25. European Union. *Presidential Decree 81/8.07.2019*. Published at the Greek Government Gazette A' 119, p. 3181 et seq.
26. Wekerle, T., Pessoa Filho, J. B., Loures da Costa, L. E. V., & Trabasso, L. G. (2017). Status and trends of smallsats and their launching vehicles – An up-to-date review. *Journal of Aerospace Technology and Management, 9*(3) Accessed from https://jatm.com.br/jatm/article/view/853
27. Palmroth, M., Tapio, J., Soucek, A., Perrels, A., Jah, M., Lönnqvist, M., Nikulainen, M., Piaulokaite, V., Seppälä, T., & Virtanen, J. (2021). Toward sustainable use of space: Economic, technological, and legal perspectives. *Space Policy, 57*, 101428. https://doi.org/10.1016/j.spacepol.2021.101428

28. Chung, G. (2018). The emergence of environmental protection clauses in the outer space treaty: A lesson from the Rio principles. In A. Froehlich (Ed.), *A Fresh view on the outer space treaty* (Studies in space policy) (Vol. 13). Springer. https://doi.org/10.1007/978-3-319-70434-0_1
29. Yang, T. (2019). The emergence of the environmental impact assessment duty as a global legal norm and general principle of law. *Hastings Law Journal, 70*, 525. Accessed from https://www.hastingslawjournal.org/wp-content/uploads/70.2-Yang.pdf
30. Mustow, S. E. (2018). Environmental impact assessment (EIA) screening and scoping of extraterrestrial exploration and development projects. *Impact Assessment and Project Appraisal, 36*(6), 467–478. https://doi.org/10.1080/14615517.2018.1500092
31. Kramer, W. R. (2014). Extraterrestrial environmental impact assessments: A foreseeable prerequisite for wise decisions regarding outer space exploration, research, and development. *Space Policy, 30*(3), 215–222. https://doi.org/10.1016/j.spacepol.2014.06.003
32. Handmer, A., & Freeland, S. (2022). The use of law to address space debris mitigation and remediation: Looking through a science and technology lens. *Journal of Air Law and Commerce, 87*, 375. https://doi.org/10.25172/jalc.87.3.2
33. Alessi, D. (2005). The Distinction between Obligations de Résultat and Obligations de Moyens and the Enforceability of Promises. *European Review of Private Law, 13*(5), 657–692. https://doi.org/10.54648/erpl2005041
34. von Der Dunk, F. (2017). Transfer of ownership in orbit: From fiction to problem. In M. Hofmann & A. Loukakis (Eds.), *Ownership of satellites: 4th Luxembourg workshop on space and satellite communication law*. Nomos Verlagsgesellschaft and Hart Publishing. Accessed from https://digitalcommons.unl.edu/cgi/viewcontent.cgi?article=1102&context=spacelaw
35. Dimopoulou, A. (2019). Is the launching state the only "appropriate state" to register a space object? Change of registry in case of change of ownership. In G. D. Kyriakopoulos & M. Manoli (Eds.), *The space treaties at crossroads*. Springer. https://doi.org/10.1007/978-3-030-01479-7_7
36. NASA. (n.d.). *5 hazards of human spaceflight*. Accessed July 2024, from https://www.nasa.gov/hrp/hazards/
37. Yates, H., & Grimaila, M. R. (2008). A systematic approach to securing our space asset. *High Frontier, the Journal for Space and Missile Professionals, 4*(2), 48–53. Accessed from https://scholar.afit.edu/facpub/160/
38. Pelton, J., Sgobba, T., & Trujillo, M. (2015). Space safety. In K.-U. Schrogl et al. (Eds.), *Handbook of space security. Policies, applications and programs*. Springer. https://doi.org/10.1007/978-1-4614-2029-3
39. EUSPA. (2024, April 29). *The EUSPA Programme. SSA*. Accessed from https://www.euspa.europa.eu/eu-space-programme/ssa
40. UNOOSA. *Convention on international liability for damage caused by space objects (Liability Convention), signed on March 29th, 1972 and entered into force in September 1st, 1972*. Accessed from https://www.unoosa.org/oosa/en/ourwork/spacelaw/treaties/introliability-convention.html
41. Reinert, A. P. (2020). Updating the liability regime in outer space: Why spacefaring companies should be internationally liable for their space objects. *William & Mary Law Review, 62*(1) Accessed from https://scholarship.law.wm.edu/wmlr/vol62/iss1/7
42. St. John, D. (2012). The trouble with Westphalia in space: The state-centric liability regime. *Denver Journal of International Law & Policy, 40*(4).
43. UNCOPUOS STSC. (2002). *Inter-Agency Space Debris coordination committee space debris mitigation guidelines*. UN Doc A/AC.105/C.1/L.260 (29 November 2002). Accessed from https://www.unoosa.org/pdf/limited/c1/AC105_C1_L260E.pdf
44. *Legislative Decree n° 670/1970 on the ratification of the treaty on principles governing the activities of states in the exploration and use of outer space, including the moon and other celestial bodies signed on 27.1.67*. Greek Government Gazette, Vol. A '208.
45. *Law n° 563/1977, on the ratification of the liability convention published in the Greek Government Gazette A '75* (pp. 595 et seq).

46. Spencer, R. (2009). State supervision of space activity. *Air Force Law Review, 63*, 75. Accessed from https://link.gale.com/apps/doc/A201030788/AONE?u=anon~c7f16181&sid=googleScholar&xid=7624d1d1
47. UNOOSA. (1961).*International co-operation in the peaceful uses of outer space, Res 1721 B (XVI), UNGAOR, UN Doc A/RES/1721(XVI)B*
48. UNOOSA. *Convention on registration of objects launched into outer space (Registration Convention), signed in January 14th, 1975 and entered into force in September 15th, 1976.* Accessed from https://www.unoosa.org/oosa/en/ourwork/spacelaw/treaties/introregistration-convention.html
49. UNOOSA. *Agreement on the rescue of astronauts, the return of Astronauts and the return of objects launched into outer space (Rescue Agreement), signed on April 22nd, 1968 and entered into force in December 3rd, 1968.* Accessed from https://www.unoosa.org/oosa/en/ourwork/spacelaw/treaties/introrescueagreement.html
50. Martin, R. J. (1980). Legal ramifications of the uncontrolled return of space objects to earth. *Journal of Air Law and Commerce, 45*(2) Accessed from https://scholar.smu.edu/jalc/vol45/iss2/5/
51. *Decision of the Minister of Digital Governance n° 34827 EΞ 2021/1.10.2021 titled "Registration of the satellite Hellas Sat 4 in the National Registry for Space Objects"*. Published at the Greek Government Gazette B' 4623/06.10.2021, p. 61149 et seq.
52. UNOOSA. *Model registration form for registration of space objects with the Secretary-General.* Accessed from https://www.unoosa.org/oosa/spaceobjectregister/resources/index.html
53. UNCOPUOS STSC. (2021). *Statements to the 945th meeting of the Scientific and Technical Subcommittee, Monday, 26 April 2021 - Item 3: General exchange of views and introduction of reports submitted on national activities (Greece)*. Accessed from https://www.unoosa.org/oosa/en/ourwork/copuos/stsc/2021/statements.html
54. UN Secretariat COPUOS. (2008). *Information furnished in conformity with the Convention on registration of objects launched into outer space note verbale dated 14 May 2008 from the Permanent Mission of Greece to the United Nations (Vienna) addressed to the Secretary-General.* 15 May 2008. UN DOC ST/SG/SER.E/INF.20. Accessed from https://digitallibrary.un.org/record/630856?ln=en&v=pdf
55. *Law n° 3131/2003 on the Ratification of the Registration convention, published in the Greek Government Gazette A '79* (pp. 1279 et seq.)
56. UN Secretariat COPUOS. (2008). *Information furnished in conformity with the convention on registration of objects launched into outer space. Note verbale dated 25 March 2004 from the permanent mission of Greece to the United Nations addressed to the Secretary-General. 29 March 2004.* UN DOC ST /SG/SER.E/446*. Accessed from https://www.unoosa.org/documents/pdf/ser446E.pdf
57. UNOOSA. *UN Secretariat, Information furnished in conformity with the convention on registration of objects launched into outer space - Note verbale dated 2 March 2022 from the Permanent Mission of Greece to the United Nations (Vienna) addressed to the Secretary-General, 19 April 2022, ST/SG/SER.E/1044.* Accessed from https://www.unoosa.org/oosa/en/osoindex/data/documents/gr/st/stsgser.e1044.html
58. UNOOSA. *Outer space objects index.* Accessed from https://www.unoosa.org/oosa/osoindex/index.jspx?lf_id=
59. Pardini, C., & Anselmo, L. (2019). Uncontrolled re-entries of spacecraft and rocket bodies: A statistical overview over the last decade. *The Journal of Space Safety Engineering, 6*, 30–47. https://doi.org/10.1016/j.jsse.2019.02.001
60. Von der Dunk, F. G. (2008). *A sleeping beauty awakens: The 1968 Rescue Agreement after forty years* (p. 29). Space, Cyber, and Telecommunications Law Program Faculty Publications. Accessed from https://digitalcommons.unl.edu/spacelaw/29

61. United Nations Office for Outer Space Affairs. (n.d.). *List of reported space objects discovered by member states. Recovery and return of objects launched into outer space*. Retrieved from https://www.unoosa.org/oosa/en/treatyimplementation/arraart-v/unlfd.html
62. UN OOSA. *Recovery and return of objects launched into outer space, list of reported space objects discovered by member states*. Accessed July 2024, from https://www.unoosa.org/oosa/en/treatyimplementation/arra-art-v/unlfd.html
63. Byers, M., Wright, E., Boley, A., & Byers, C. (2022). Unnecessary risks created by uncontrolled rocket reentries. *Nature Astronomy, 6*. https://doi.org/10.1038/s41550-022-01718-8

Chapter 6
Conclusive Remarks

The present study attempted to identify and present the principal developments that have taken place until now, in a field that may be regarded as *Greek space law*. The major advances described therein occurred essentially because space activities, initiated at the national level but conducted—as a general rule—in collaboration with the European and international space-related bodies and institutions, are burgeoning thanks to a new generation of space products. Such a trend allows Greece to be considered as an emerging space-faring country, able to contribute to further strengthening the space sector both at the legislative and policy levels (this was the case, for instance, when the Greek State acceded—on 9 February 2024—to the Artemis Accords), in addition to the scientific domain where key milestones have been achieved, as briefly presented below.

Indeed, recent steps towards advancing Greek space research and enhancing space-related infrastructures demonstrate Greece's growing presence in the space sector. For instance, the strategic cooperation between the National Observatory of Athens (NOA) and the European Space Agency (ESA), that was ratified through a memorandum of cooperation, and the successful 2020 evaluation of NOA's Aristarchus 2.3 m telescope by ESA mark significant progress. This followed a period of tests, preparatory work and the successful establishment of regular links with the AlphaSat satellite, which resulted in its selection as the first European Optical Ground Station (OGS) for optical satellite communications under the ARTES/Scylight (SeCure and Laser communication Technology) program of ESA. As a result of the above, the Aristarchos OGS will be utilised also under the EuroQCI development programme (note that NOA's Aristarchos Telescope is the largest telescope of southeastern Europe and the second largest in mainland Europe), while two more Greek telescopes have been selected by ESA as OGS in the north (Cholomondas telescope) and the south (Skinakas telescope) [1].

Thus, the ARTES/Scylight/HydRON/SAGA optical and quantum communications programs of ESA aim in the development of Optical Communications, Photonics and Quantum technologies that will enable satellite communications without bandwidth limitations, licensing requirements, interference, jamming or link

A. Koskina, M. Plionis, *Greek Space Law*, Space and Society,
https://doi.org/10.1007/978-3-031-72813-6_6

detection by third parties, with small size, weight and power requirements on board the satellite. As a result, said development projects will strengthen the European scientific and technological capacities in optical/quantum communications and cyber security as well as in deep-space optical communications.

In this respect, ESA and NOA are also closely collaborating in the emblematic *Psyche* NASA mission, which is on its way to the asteroid 16-Psyche, with NOA being a key partner, according to a relevant ESA announcement. This mission will serve as a pioneering test-bed for deep-space optical communications, aiming for the first time to achieve optical communication over a distance of about 300,000,000 km, equivalent to twice the Sun-Earth distance. This will take place in the summer of 2025 and, if successful, will constitute a historic achievement, namely the deepest optical-communications link in space. In this pioneering experiment NOA participates with its two telescopes, located in the Peloponnese, the *Aristarchus* telescope at Helmos mountain and the Kryoneri 1.2 m telescope near Korinth. It is worth mentioning that NOA's telescopes are the only European telescopes participating in this experiment.

The above capabilities developed by NOA led to the close cooperation between NOA and the private company *HellasSat* which is the national satellite provider and contractor of the Greek State in the field of optics and quantum communications. This collaboration created the conditions for the recent Greek-French partnership between *HellasSat*/NOA and Thales Alenia Space/CNES under the overall supervision of ESA. Specifically, *HellasSat* agreed with Thales Alenia Space to work on the development of an optical communications payload that will be financed by the French National Space Agency (CNES) and other agencies. The optical payload will be designed, developed and installed on the new *Hellas Sat 5* satellite mission, which will operate from geostationary orbit 39 degrees east. The optical payload operated by the *Hellas Sat 5* satellite will be connected to the optical ground stations of the National Observatory of Athens (NOA) and to optical stations in France, to support optical communications at a European Union level.

Other important initiatives, led by research and academic Greek institutions, have resulted in specific collaborations with ESA on a variety of missions, including the participation of NOA in the *Cloud, Aerosol and Radiation Explorer* (EarthCARE) mission, a joint ESA-JAXA (Japan Aerospace Exploration Agency) mission, where the Greek side together with the German Aerospace Center DLR, the LATMOS Laboratory of the University of Paris and the British research laboratory FAAM will implement, on behalf of ESA, the ACROSS experiment, with the primary objective of calibrating and validating the EarthCARE observations. To this end NOA will utilize its novel research infrastructures of the PANGEA Climate-Change and Geophysics Observatory in Antikythira.

The steps taken by Greece in the field of space are by no means limited to the space research activities presented above, or to the development and implementation of projects for the utilization of space resources addressed at any point in this study. Be that as it may, by mentioning here a few of these, one could argue that the overall dynamic of the Greek space sector as a whole has been demonstrated adequately.

References

1. Hellenic Republic, Ministry of Digital Governance - General Secretariat of Telecommunications and Post.(2021). *Request for information (RFI): Outline concept for end-to-end small satellites multipurpose solutions inresponse to national and European needs* (Date of issue: 28.06.2021). Retrieved from https://mindigital.gr/wpcontent/uploads/2021/09/20210626-GR-RFI-microsatellites.docx

Annex

Translation of the Greek Space Law[1]

Law n° 4508/2017—Greek Gov. Gazette 200/A/22-12-2107

"Authorisation of Space Activities—Registration in the National Register of Space Objects—Establishment of the Hellenic Space Agency and Other Provisions"

Chapter A

Authorisation of Space Activities—Registration in the National Register of Space Objects

Article 1: Subject Matter—Scope

1. The provisions laid down in this law regulate:

   (a) The terms and procedure of authorization, liability for damage caused by space objects and supervision over space activities,
   (b) The setting up and maintenance of the National Register for recording and registering space objects.

[1] This is a *verbatim translation* of the Greek space law, which was made for the purposes of this book and reflects the views of the authors only.

A. Koskina, M. Plionis, *Greek Space Law*, Space and Society,
https://doi.org/10.1007/978-3-031-72813-6

2. The provisions of this law shall apply to space activities:
   (a) Carried out within the Greek territory,
   (b) Carried out outside the Greek territory, in cases where infrastructure, personnel, movable or immovable property being used belong(s) to the Greek State or fall under its jurisdiction,
   (c) Irrespective of their location, if they are carried out by natural persons of Greek nationality or legal entities established on the Greek territory and only if this is provided for by an International Agreement or Treaty,
   (d) Carried out by natural or legal persons, domestic and foreign, within the Greek territory or for which infrastructure, personnel, movable or immovable property of the Greek State or falling under its jurisdiction and placed under the responsibility of the Greek State for being the "Launching State" in accordance with International Law is being used.
3. In any case the Greek State is held responsible in accordance with International Treaties on Outer Space.
4. In case other States meet the requirements as Launching States pursuant to paragraph 2 of Article II of the Convention on Registration of Objects Launched into Outer Space, which was ratified by Law n° 3131/2003 (A' 79), the Minister of Digital Policy, Telecommunications and Information is competent to decide on the registration of a space object for Greece.
5. Space activities related to space objects listed in the space objects registry of another State and falling under that State's international responsibility for damage, in accordance with International Law, shall be exempted from the provisions of this Law. In this case, the operator shall without any delay inform the Minister of Digital Policy, Telecommunications and Information about the space activity referred to above.

Article 2: Definitions

For the purposes of this Law the following definitions shall apply:

1. "Space object":
   (α) Any object launched or intended to be launched into orbit around the Earth or to a destination beyond Earth orbit, in outer space,
   (β) Any device or system intended to be used for launching an object, as referred to in point α', even in case such a device is not carrying any payload, operating with the sole aim of developing and validating its proper functioning,
   (γ) Any removable part of an object, as referred to in points α' and β',
2. "Operator conducting space activities" or "operator": the natural or legal person carrying out or undertaking to carry out space activities, ensuring in person or together with other operators the effective control of the space object, following a relevant agreement or authorization. If a space object cannot be maneuvered after being put in orbit, the operator shall be the person giving the order to place the space object in orbit,

3. "Effective control": the activation of the means of control and command, or remote control and command, and of any monitoring necessary to carry out launching, flying and maneuvering activities of one or more than one space objects,
4. "Manufacturer": any person who participates or has participated in the development, manufacture or assembly of all or part of a space object,
5. "Flight operation and guidance": any procedure concerning the placement in orbit, flight, navigation, trajectory, or launching of a space object in outer space such as, mainly, the choice, review, or correction of its course or orbit trajectory,
6. "Space activities": launching, flight-related and maneuvering activities, as well as those related to the return of space objects,
7. "Outer Space Treaty": The Treaty on Principles Governing the Activities of States in the Exploration and Use of Outer Space, including the Moon and Other Celestial Bodies, signed on January 27th, 1967, and ratified by Greece on 3 October, 1970 (Legislative Decree n° 670/1970, A' 208),
8. "Treaty on International Space Liability[2]": The Convention on International Liability for Damage Caused by Space Objects, signed on 29 March 1972, and ratified by Greece on 12 March 1977 (Legislative Decree n° 563/1977, A' 75),
9. "Convention on the registration of Space Objects[3]": the Convention on Registration of Objects Launched into Outer Space, signed on 14 January 1975, and ratified by Greece on 4 April 2003 (Law n° 3131/2003, A' 79),
10. "Agreement for the Rescue of Astronauts and the Return of Space Objects[4]": the Agreement on the Rescue of Astronauts, the Return of Astronauts, and the Return of Objects Launched into Outer Space, signed on 22 April 1968, and ratified by Greece on 20 November 1974 (Legislative Decree n° 189/1974, A' 350),
11. "Launching State":

    (α) A state launching or procuring the launch of a space object,
    (β) A state from whose territory or facility a space object is launched, according to the definition provided in Article VII of the Outer Space Treaty, in the first Article of the Liability Convention or in the first Article of the Registration Convention,

12. "Damage": any type of damage as it is defined in the first article of the Liability Convention,
13. "Transfer of authorized space activities": the transfer of authorization and of rights related to space activities in general,
14. "State of Registry": a launching State on the registry of which a space object is listed, in accordance with the first Article of the Registration Convention.

---

[2]Hereafter, also *Liability Convention*.

[3]Hereafter, also *Registration Convention*.

[4]Hereafter, also *Rescue Agreement*.

Article 3: Authorization Requirement

1. To engage in space activities (namely the launching, operating, maneuvering and return of space objects), a prior authorization is required.
2. The authorization is granted to the operator upon their request by decision of the Minister of Digital Policy, Telecommunications and Information, applies each time to a specific space activity and is personal.
3. Space activities must be conducted in accordance with the principles of International Law and International Treaties, Conventions and Agreements, to which Greece is a State Party.

Article 4: Authorization Conditions

1. The authorization is granted for an undetermined or determined period and shall be renewable, when the following conditions are fulfilled:

   (α) The operator has the necessary credibility, competence, and experience in the conduct of space activities,
   (β) The space activity does not constitute a threat to national security, public policy, the security of persons and property, and public health,
   (γ) The space activity does not conflict with Greece's obligations under International Law or the strategic plans reflecting Greek foreign policy,
   (δ) Appropriate provision has been made for space debris or residues mitigation and management in accordance with the technological state of the art and international practices,
   (ε) The space activity does not cause pollution of outer space or celestial bodies or negative changes to the environment,
   (στ) The operator satisfies the requirements of the International Telecommunication Union (ITU) as regards orbital positions and their corresponding radiofrequencies,
   (ζ) The operator satisfies the requirements for insurance cover provided for in Article 11,
   (η) The operator has taken appropriate measures for a smooth and safe conclusion of the space activity,

2. By decision of the Minister of Digital Policy, Telecommunications and Information, additional specific requirements may be adopted for the authorization, to ensure the security of persons and property, protect the environment, ensure optimal use of the atmosphere and outer space, preserve the Greek State's strategic, geopolitical and financial interests, and comply with the Greek State's obligations under international law.
3. The Minister of Digital Policy, Telecommunications and Information, may impose on any decision granting an authorization additional or any specific conditions that he considers useful, on a case-by-case basis. More specifically, the Minister may ask for technical assistance from third parties, such as Greek, European, and international organizations and agencies, experts, qualified advisers, and scientists, set conditions for the space activities' location or for

the operator's main establishment location or the provision of insurance for the space activities and space objects.

4. The Minister of Digital Policy, Telecommunications and Information may change, by a duly justified decision, the specific conditions applying to an authorized space activity, as well as to their duration.

Article 5: Content of the Dossier and Procedure

1. The application for a space activity authorization shall be submitted by the operator and co-signed by any participants in the space activity submitted for authorization, as well as by any joint owner of the space object, to the Minister of Digital Policy, Telecommunications and Information, accompanied by a fee of six thousand (6.000) euros. The Minister of Digital Policy, Telecommunications and Information may exempt from payment of the fee space activities intended exclusively or in part for public purposes or for exclusively non-commercial research, experimental and educational purposes.
2. The application shall be accompanied by:

   (α) Full identification details of the operator and of its legal form (legal documents),
   (β) A presentation of their activities in general, and of the operator's guarantees regarding the viability of the space activity submitted for authorization,
   (γ) A complete and accurate description of the space activity and of the space object for which authorization is sought,
   (δ) Full details of the manufacturer of the space object,
   (ε) A declaration on the operator's honor regarding the compliance with environmental requirements and practices, as well as with the national and international environmental legislation in force, as referred to in Article 6, together with a comprehensive presentation of the environmental protection forecasting and measures,
   (στ) Full identification details of the natural and legal persons on behalf of which the space activities will be carried out,
   (ζ) Full contact details of the persons who have been designated by the operator as contact points,
   (η) Contract of insurance cover, in accordance with Article 11,
   (θ) The required elements to register in the National Register for Space Objects.

   The operator may include in the application any other information or supporting document he considers relevant in the justification of his request.

3. The list of supporting documents of paragraph 2 may be amended by a decision of the Minister of Digital Policy, Telecommunications and Information.
4. The Minister of Digital Policy, Telecommunications and Information may ask from the operator for any additional information which he considers necessary for the examination of the request. Any failure to provide it within a reasonable period of time that will be set by the Minister is sufficient grounds to reject the application.

5. In examining the request, the Minister of Digital Policy, Telecommunications and Information may seek the opinion of experts, natural or legal persons, whether governed by public or private law, with many years' international experience in the field of space, regarding the credibility, expertise and experience of the operator, the manufacturer's credibility in the relevant fields and their ability to comply with existing best practices and rules, as well as on the operator's solvency. The experts appointed by the Minister of Digital Policy, Telecommunications and Information have access to the facilities, buildings and materials that will be used by the operator for the purpose of carrying out the space activities, and will ensure the secrecy and confidentiality of the procedure. Any denial of access is sufficient grounds to reject the application.
6. The application dossier is communicated to the Ministries of Economy and Development; National Defense; Education, Research and Religious Affairs; Foreign Affairs and Environment and Energy, which may raise their objections within a period of forty (40) days, otherwise consent of any Ministry concerned will be presumed.

Article 6: Environmental Impact Assessment

1. For each space activity, the space operator is required to submit, at successive stages, to the Minister of Digital Policy, Telecommunications and Information, reports on the impact of the space object on the environment. The Minister of Digital Policy, Telecommunications and Information together with any other competent Minister in each specific case may designate one or more experts for reviewing these assessments.
2. The initial assessment shall be made prior to the granting of authorization for the purpose of assessing the potential environmental impacts on the Earth or outer space due to the launch or operation of the space object.
3. After the launching of the space object or during its operation an interim report shall be established presenting the real impacts of the space activities on the environment of the Earth or outer space.
4. Following the re-entry of the space object into the Earth's atmosphere, after the completion or the cessation of the space object's operation, a final report shall be established and submitted.
5. By a joint decision, the Ministers of Digital Policy, Telecommunications and Information, and Environment and Energy determine the content of the reports, the requirements and technical standards, that the operator's activity has to comply with. Pending the adoption of the above-mentioned decision, the existing national, European and international standards will apply, as well as the related good practices. Fulfillment of the legal requirements, as regards the environmental impacts and the relevant authorization conditions, may be presumed in case of compliance with the relevant standards.
6. Costs of the reports mentioned above will be borne by the operator.
7. The operator signs a declaration of honor regarding his compliance with the aforementioned standards and practices, and with the environmental requirements in general, which declaration binds in any case the operator to the accuracy and completeness of the reports.

Article 7: Granting of Authorization

1. The Minister of Digital Policy, Telecommunications and Information decides on the approval or rejection of the application, in accordance with Articles 3 to 6, within a period of three (3) months from its submission. The authorization shall determine the space activity for which the permit is issued and may contain specific conditions.
2. If the Minister of Digital Policy, Telecommunications and Information requires from the operator additional information, in accordance with paragraph 4 of Article 5 or, if he asks for further examination of the application, the time period mentioned above may be extended for three months.
3. If no information is received on expiry of the time period established in paragraphs 1 and 2 of this Article, the application is deemed rejected.
4. The decision of authorization is published in the Greek Government Gazette.

Article 8: Control of Activities

1. The Minister of Digital Policy, Telecommunications and Information may designate experts, natural or legal person, whether governed by public or by private law, with many years' international experience on space matters or/and on aerospace systems, for the purpose of supervising the space activities carried out by the operator. The operator shall take all necessary measures to facilitate the conduct of inspections and controls of the space activities carried out, at any time.
2. To carry out the inspections and controls of the space activities under examination, experts may have access to any information or material in the operator's possession, up-to-date information and data generated by space activities, as well as to installations that have, directly or indirectly, been used for space activities.
3. All the information obtained during the inspections or controls carried out by experts designated for that purpose, shall be considered confidential.
4. The operator's refusal to grant the experts such access is sufficient ground for the suspension or the withdrawing of the authorization.

Article 9: Withdrawal or Suspension of Authorization

1. The authorization may be withdrawn or suspended for a certain period of time by decision of the Minister of Digital Policy, Telecommunications and Information in the following cases:

   (α) If one of the general or particular conditions of authorization is not being complied with,
   (β) In case of any infringement of the rules governing the granting of authorization,
   (γ) For reasons of public order, security and health.

2. In the cases referred to above, prior to adopting any decision, the operator is invited to provide explanations and to propose measures with a view to ending the infringement or remedy the situation, within a reasonable time-limit that will be established by the Minister. In cases of emergency, no prior hearing is required.

3. In case of a withdrawal or suspension of the authorization and following the operator's written request, the Minister of Digital Policy, Telecommunications and Information may adopt provisional measures for the management of the ongoing space activities, to assist him in fulfilling his contractual obligations. Any request by the operator shall be made without undue delay on his becoming aware of a decision to withdraw or suspend the authorization.
4. In case the authorization is withdrawn or suspended after the launching of the space object in outer space, the Minister of Digital Policy, Telecommunications and Information shall take the necessary measures to ensure the space activity's temporary continuation or safe cessation, to guarantee the safety of space activities, as well as to ensure the protection of assets and of the environment. For this reason, the Minister of Digital Policy, Telecommunications and Information may give any instructions to the operator or to third parties, have recourse to the services of a third party, or transfer the space activities to another operator, so as to guarantee the continuity of space flight and guidance and, where necessary, to de-orbit, deactivate or destroy the space object.
5. Decisions on the withdrawal and suspension of the authorization are published in the Greek Government Gazette.

Article 10: Transfer of Activities

1. The transfer of activities is only allowed following an approval of the Minister of Digital Policy, Telecommunications and Information.
2. Any request for the approval of a space activities transfer shall be presented jointly by the space transferor and the space transferee.
3. The rules regulating the initial authorization procedure shall apply accordingly to the procedure for the approval of a space activity transfer.
4. The Minister of Digital Policy, Telecommunications and Information may, on the occasion of any space activities transfer, impose additional conditions on both the transferor and the transferee.
5. When the transferee is not located or active in Greece, the Minister of Digital Policy, Telecommunications and Information may refuse to approve a space activities' transfer, in case there is no signed agreement with the transferee's country to provide for the indemnification of the Greek State against claims based on international responsibility or legal actions for damage caused.

Article 11: International Liability for Damage

1. To cover his liability for damage, the operator must take out an insurance coverage for the amount of 60.000.000 Euros, which may be adapted depending upon the size, value or other operating parameters of the space object by a decision of the Minister of Digital Policy, Telecommunications and Information further to the operator's request, which must be made prior to the submission of the supporting documents mentioned in paragraph 2 of Article 5. If the space activity is conducted for the benefit of the public interest, the Minister of Digital Policy, Telecommunications and Information may, by decision, establish a lower amount or exempt the operator from the obligation to take out an insurance, taking into account the risks inherent in space activity. Space activities are in the

public interest if they serve the purposes of national defense and security, science, research or education. Insurance coverage is not necessary if the operator is the Greek State.

2. Where the Greek State is internationally liable to pay compensation, pursuant to Article VII of the Outer Space Treaty, the provisions of the Convention on International Liability for Damage caused by Space Objects or this law, it shall be allowed to exercise recourse against the operator involved, up to the amount set out in paragraph 3.
3. Damage assessment between the State and the operator shall be made as following:

   (α) In the case of paragraph 2, where the damage has been caused to a third country or to third-country nationals, the damage assessment shall be conducted between the Greek State and the State representing the victim, in line with the Convention on International Liability for Damage caused by Space Objects. The operator, or the person designated for this purpose by him, may participate in the discussions and damage assessment procedures conducted by the representatives of the States involved, in order to defend his interests,

   (β) In the case of paragraph 2, where the damage has been caused to Greek nationals the assessment shall be conducted by a committee of three (3) experts, one of whom shall be appointed by each party and the third one shall be chosen by common agreement of the parties. The Minister of Digital Policy, Telecommunications and Information may impose the ex-ante designation of the experts as one of the conditions for the issuance of the authorization.

4. The right of the Greek State to bring a claim against another launching or co-launching State, in accordance with paragraph 2 of Article V of the Convention on International Liability for Damage caused by Space Objects, does not prevent the application of the present article and does not, in any event, constitute a condition that must be met before the Greek State can take action against the operator
5. The Greek State is entitled to file a claim against the operator's insurer, up to the amount determined in accordance with paragraph 3, and it may be designated as a joint beneficiary to receive any compensation paid by the insurance or any other guarantee.
6. The operator is required annually to submit a copy of the insurance contract or a certificate from the insurance company indicating that the insurance contracted by the operator is in force.

Article 12: Obligation to Inform

The operator shall inform without delay the Minister of Digital Policy, Telecommunications and Information of any kind of handling, malfunction or anomaly of the space object, as well as general loss of control, which may put at risk people on Earth, airplanes in flight or other space objects or cause any damage, as well as for any change in the information contained in the application for authorization.

Article 13: Return of the Space Object

1. Without prejudice to the measures concerning the safety and protection of assets and persons, each space object found or identified on Greek territory or on a territory under Greek jurisdiction, shall be returned without delay to the competent authorities, who shall immediately notify the Minister of Digital Policy, Telecommunications and Information, to enable them to arrange for the return of said object to the State of registry, in accordance with the Rescue Agreement.
2. Where an investigation is required for the purpose of determining the State on the registry of which the space object is listed or of the launching State, the Minister of Digital Policy, Telecommunications and Information shall take any measure necessary for the preservation of the object or objects found or identified.
3. When the object is returned to the competent authority or, if required, before its return, any necessary measures shall be taken for the protection of victims' rights, as well as for the property that has been damaged by the space object.

Article 14: Coverage of Application Costs

When, in accordance with this law, the Minister of Digital Policy, Telecommunications and Information has recourse to the services of experts, costs of the services, and any expenses related to their services shall be borne by the operator. In case the operator does not cover the costs, his application shall be rejected; if he has already been granted a license, the Minister of Digital Policy, Telecommunications and Information may suspend the authorization.

Article 15: Penalties

1. Any natural person and any duly empowered representative of a legal person, carrying on space activities without an authorization, is liable to a punishment of up to two (2) years of imprisonment, and a fine of between 30.000 and 300.000 Euros.
2. In case that the Greek State is internationally responsible for the unauthorized space activities, its right to seek indemnification from the natural or legal person involved is hereby established, up to the compensation amount determined in accordance with paragraph 3 of Article 11.
3. Anyone who, at the application stage and until the authorization expires, intentionally provides any inaccurate or incomplete information to the Minister of Digital Policy, Telecommunications and Information in reference to space activities, is punishable by a fine of 10.000 to 100.000 Euros.

Article 16: Transitional Provisions

1. Space activities which are already under way at the time of entry into force of this law may be continued for a period of twelve (12) months as from this date, without any requirement for authorization. No transfer can take place during that period.
2. The operator must, within twelve (12) months of the entry into force of this law, submit information on the space activities being carried out, as provided for in Articles 5 and 6, to the Minister of Digital Policy, Telecommunications and Information. In such a case, Article 15 will not apply to the ongoing activities.

The operator provides, in accordance with the circumstances particular to each case, all the relevant documents allowing verification of the conditions set out in paragraph 2 of Article 5.

3. Space activities which are included in specific contracts or agreements with the Greek State and have received the relevant authorizations, are considered as having been licensed and are subject to the corresponding obligations and conditions of this law.

Article 17: National Register for Space Objects

1. A National Register for Space Objects is set up and kept at the Ministry of Digital Policy, Telecommunications and Information, listing the space objects related to authorized activities, for which Greece is, whether alone or together with another State, the launching State, unless the registration is made by another State or International Organization, in accordance with the Convention on Registration of Objects Launched into Outer Space.
2. For the registration of space objects the following shall apply:

(α) A unique national registration and listing number shall be attributed to each object, following a decision by the Minister of Digital Policy, Telecommunications and Information.

(β) The information to be recorded shall include the elements referred to in Article IV of the Convention on Registration of Objects Launched into Outer Space, as well as those mentioned in the application for a space object registration at the UN Register.

3. Additionally to the information referred to in case β' of paragraph 2, the following shall also be recorded confidentially:

   (α) The manufacturer of the space object,
   (β) The company launching the space object,
   (γ) The operator and the owner of the space object,
   (δ) The component elements and parts inside the space object,
   (ε) The ITU filings in accordance with which the object is operating,
   (στ) The authorization to engage in space activities which has been issued to the operator together with related supporting documentation.

4. The operator shall notify without any delay the Minister of Digital Policy, Telecommunications and Information of the information referred to in case β' of paragraph 2 and in paragraph 3. In the event of any change in the information communicated, an additional registration will be made, under the responsibility of the operator and within thirty (30) days of the date on which he became aware of the change. If the operator does not communicate such changes within the prescribed period, the authorization may be revoked or suspended.
5. The registration shall take place after the time of launching of the space object and following a written notification to the Minister of Digital Policy, Telecommunications and Information by the operator about the launch time, the launch and placing into operation of the space object and provided the authorization conditions are met.

6. By a decision of the Minister of Digital Policy, Telecommunications and Information, space objects relating to space activities in progress at the time of entry into force of this law shall also be listed in the Register.

Chapter B'

Establishment of the Hellenic Space Agency

Article 18: Establishment of a Public Limited Company Under the Name "HELLENIC SPACE AGENCY S.A."

1. A public limited company named "HELLENIC SPACE AGENCY S.A." and abbreviated as the H.S.A. is hereby established, hereinafter "the Company". In its international relations, the name of the company will be translated in English as "HELLENIC SPACE AGENCY". The company's headquarters are located in the municipality of Athens, unless another municipality is chosen for the location of its headquarters following a decision of the Minister of Digital Policy, Telecommunications and Information.
2. The company is supervised by the Minister of Digital Policy, Telecommunications and Information, operates to the benefit of the public interest under market economy principles, and is governed by the provisions of this law, of law n° 3429/2005 (A' 314) as currently in force, and additionally by the provisions of the implementing decree n° 2190/1920 (A' 76) for the matters not being explicitly regulated.
3. The term of the company is set at thirty (30) years, starting from the day of publication of the Ministerial Decision establishing the statutes for the Company referred to in paragraph 6, and may be amended or extended as provided for in the statutes.
4. The objective of the company is:

   (α) The development of a proposal for a strategy in the field of space and the establishment of a rolling action-plan of the space strategy in collaboration with the academic and research community, the public and the private sector, to set out the objectives, areas of cooperation and procedures needed for achieving these objectives,

   (β) The establishment of cooperation, coordination, support and mobilization of operators, administrations, and legal persons governed by public or private law in order to promote, disseminate and make full use of the Greek space strategy, including participation in European and international organizations and clusters which are active in space matters and programs,

   (γ) The promotion and participation as a coordinator of public bodies in projects and programs, including management of national programs and projects in such space fields as scientific research, technology, telecommunications, security, economy, environment, rural development, transports, e-government and trade,

(δ) The participation in European and international organizations, initiatives, fora and activities on space matters, the coordination of national representatives in the field of space, the development of European and international collaborations, the exchange and consolidation of knowledge, innovation and actions in order to maximize the contributory participation of the country in European and international space organizations, and in European or international space programs, including its participation in the formulation of programs aimed at meeting national, scientific, operational and economic development needs in which space plays a primary role,

(ε) The contribution to the development and enhancement of technology and space applications, services and ground-based infrastructures to the benefit of the domestic industry and research, as well as the support in the design of satellites, satellite systems, materials and equipment, including remote sensing activities, through the full use of the country's memberships in international and European organizations and programs, and the returns of all kinds to which the country is entitled,

(στ) The cooperation with competent bodies and administrations for the continuous training of students, young researchers and public officials in space matters, including the elaboration and participation in the implementation of actions aimed at developing understanding of space activities and applications,

(ζ) The promotion of matters relating to standardization and the transfer of know-how and best practices in the field of space applications and services,

(η) The experimental or commercial exploitation of the State's rights in satellite and space objects, including the exploitation of related services, resources and assets, movable and immovable, following a relevant rights assignment or a permission by the Minister of Digital Policy, Telecommunications and Information,

(θ) Assistance to the State in reference to satellite and space matters, including the provision of advisory services and the undertaking of studies for the Ministry of Digital Policy, Telecommunications and Information regarding the management of the State's rights and obligations in space, the registration and evaluation of space objects, as well as the development of satellite systems.

5. The share capital of the Company is set at the amount of sixty thousand (60.000) Euros, which is divided into sixty (60) ordinary registered shares for a nominal value of one thousand (1.000) Euros each. The Company's share capital shall be paid for in full by the Greek State. For the exercise of its shareholder's rights, the Greek State shall be represented at the General Meeting by the Ministers of Digital Policy, Telecommunications and Information, and the Minister of Finance, or by persons authorized by them.
6. Following a decision of the Minister of Digital Policy, Telecommunications and Information, published in the Government Gazette, the Company's statutes will be established, to regulate matters set out in the law on limited liability

companies and will be published in the General Commercial Registry (GEMI). The company statutes, and the provisions which are regulated by the content of the present law as well, may be amended and codified by a decision of the General Meeting. To achieve its statutory objectives, following an approval by the Minister of Digital Policy, Telecommunications and Information, the Company may set up subsidiaries and branches, and cooperate with Greek or foreign companies and bodies active in the field of space, by setting up joint ventures with them, participating in them as an associate or shareholder, or by creating with them new companies or holding companies that would pursue identical or similar purposes.

7. The company's resources stem from:

    (α) The exploitation of the State's rights on space objects, satellites, orbital positions and associated frequencies allocated to the Company by the Minister of Digital Policy, Telecommunications and Information, as well as from contracts or programs carried out in the course of experimental or economic activity,
    (β) Sponsorships, revenues and financing from State-owned undertakings, companies, bodies of all kinds, EU funds and other international organizations,
    (γ) The provision of services and the transfer of usage rights to scientific and research works,
    (δ) Donations, inheritances, legacies, revenues from interests on deposits, income derived from the exploitation of its property and any other type of income from legal sources,
    (ε) Sponsorships from individuals, institutions and companies.

8. The Company's resources are deposited in a current account kept in a domestic credit institution and managed by the Executive Director, after relevant approval by the Administrative Board. The withdrawal, allocation or payment of any expenditure from the Company's account below a predetermined limit, to be laid down in the Internal Regulations, and for daily operating costs, shall be made following an order made by the Company's Executive Director or a person legally authorized by him.
9. Part of the positive results, income minus expenditure, following the close of each financial year of the Company shall be attributed to the regular budget as a government revenue. A joint decision of the Ministers of Digital Policy, Telecommunications and Information, and Finance, shall determine the attributable percentage, the relevant procedure and any other related matter.
10. By a joint decision of the Ministers of Digital Policy, Telecommunications and Information; Finance, Economy and Development and the competent Minister in each case, the Company may be subsidized by the Program of Public Investment (PPI) or from the ordinary Budget for the realization of its purpose. In any case, EU legislation on State aids shall apply, with the exception of military applications and national security issues.
11. As regards space programs and space applications serving national defense purposes, the Ministry of National Defense keeps administrative and

operational autonomy from the Company, and may participate when convened in the meetings of the Company's Administrative Board without a voting right, in case an issue related to national defense and security is being discussed.

Article 19: The Company's Governing Bodies

1. The Company's governing bodies are the General Meeting and the Administrative Board, composed of the Chairman, the Executive Director and its members. Until the setting up, organization and recruitment of such staff as is necessary for the Company, which may not exceed two (2) years from the publication of the Statutes, the Minister of Digital Policy, Telecommunications and Information shall, by decision, appoint the first Administrative Board and regulate matters relating to its functioning.
2. The highest governing body of the Company is the General Meeting, which decides on all matters provided for in the Statutes and in the codified legislation n° 2190/1920, with the exception of matters that have to be decided upon in accordance with this law.
3. The executive management bodies of the Company are the Administrative Board and the Executive Director, whose responsibilities, means and procedure for decision-making, as well as any other relevant issue are established and detailed further in the Company's Statutes, according to the current legislation. The Executive Director is the head of the Company's Directorates and adopts the necessary implementing acts, as set out in the Company's Statutes.
4. The Administrative Board comprises seven (7) members, namely:

   (α) Chairman,
   (β) Executive Director,
   (γ) Two members, designated by the Minister of Digital Policy, Telecommunications and Information,
   (δ) One member, designated by the Minister of Finance,
   (ε) One member, designated by the Minister of Education, Research and Religious Affairs and
   (στ) One member, designated as a representative of employees.

   In case that the above do not propose any candidates to be appointed as Administrative Board members, with their respective alternates, within ten (10) days of the notification of the relevant request, the Minister of Digital Policy, Telecommunications and Information shall designate said members of the Administrative Board based on his own judgement.

5. The Administrative Board shall be appointed by the General Meeting for a 5-year mandate, and is responsible for the administration of the Company and the achievement of its statutory objectives. The Administrative Board shall decide on all matters related to the management of the Company, except for those matters which, in accordance with this law, are within the competence of the General Meeting. The Administrative Board of the Company may by its decision entrust the handling of particular issues to experts in the field of space, who are not members of the regular staff of the Company and have specialized

knowledge and experience regarding matters entrusted to them, while their remuneration shall be determined by the above decision.

6. The Chairman and the Executive Director are members of the Administrative Board of the Company, they carry out the duties assigned to them and are appointed by a decision of the General Meeting for a period of five (5) years, after announcement of the position and following a decision by the Ministers of Digital Policy, Telecommunications and Information, and Finance. The position announcement lays down the procedure for the selection, specifies the necessary qualifications, terms of employment and conditions of appointment, and may establish additional qualifications to the ones required above. The qualifications required to access the position of Chairman are: a degree from a domestic or foreign university, a master's degree preferably on a related topic, a multi-annual international experience in the field of space, experience in the administration and organization, and excellent knowledge of at least English language, whereas very good command of additional languages is desirable. The qualifications required to access the position of Executive Director are: a domestic or foreign university-level degree, a master's degree preferably on a related topic, at least five years of experience in the field of space and in the fields of administration and organization, and excellent knowledge of at least English language, whereas very good command of additional languages is desirable.
7. The members of the Administration Board must be widely renowned, possess scientific competence and professional experience, a high level of expertise and an extensive knowledge able to meet the requirements of the post or of the areas in which the Company operates. Particular qualifications and prohibitions or incompatibilities concerning the members of the Administration Board shall be determined in the Company's Statutes.

Article 20: Rules of Procedure

1. Rules, drawn up by the Administrative Board of the Company in accordance with Article 4 of Law n° 3429/2005 regulate matters of internal governance, security and staff classification, personal data and classified information protection, as well as control of the company. A decision of the Administrative Board of the Company shall lay down the conditions and procedures for the award and implementation of contracts for the supply of goods, provision of services, carrying out of studies, the award and implementation of public work contracts, property purchases, exchanges and sales of movable and immovable assets, rental of real estate, rentals and concessions to use as well as any other contract or property right on property belonging to the Company.
2. The Rules of Procedure regulate in particular the organization, functioning, structure and competences of the Company's services, the number of staff positions for each category in each area of specialization and their allocation between administrative units in the Company, the number of qualified scientists, including the number of legal counsellors, lawyers and special legal assistants, related to the Company by a remunerated mandate agreement, the necessary qualifications, the conditions for recruitment and employment and for career

and salary development applying to the qualified scientists, pursuant to the Staff Regulations, the managers' assessment procedures, and the disciplinary rules applying to the staff as well as any other relevant matter.

Article 21: Staff Matters

1. The Company shall recruit staff on the basis of a private-law employment contract for a fixed or an open-ended term, a remunerated mandate agreement or a works contract, according to the legal provisions in force.

   The Company may, by way of derogation from the relevant provisions, recruit up to ten (10) qualified scientists with renowned expertise to assist it in the achievement of its objectives and to address the particular needs of the State, in controlling space activities and overseeing space objects, by analogous application of paragraph 6 of Article 13 of Law n° 3429/2005.
2. At the beginning of the Company's operation, up to six (6) officials from the Ministry of Digital Policy, Telecommunications and Information may be assigned, by a decision of the Minister of Digital Policy, Telecommunications and Information, to the Company, in derogation from the relevant provisions and following a request from the parties concerned, either on a permanent basis or under a private-law employment contract for an unlimited term. The same decision regulates the conditions and any other issue relating to the assignment, except for salary matters falling under the scope of Article 23 of Law n° 4354/2015 (A' 176). Following a decision from the Minister of Digital Policy, Telecommunications and Information, in derogation from the relevant provisions, the immediate and urgent staffing needs of the Company can also be temporarily met by a transfer or a parallel exercise of duties from the existing qualified personnel of the Ministry of Digital Policy, Telecommunications and Information to the benefit of the Company. The same decision shall determine the specific employment conditions governing the temporary exercise of functions of these personnel until the completion of the formation and staffing of the Company, as well as all necessary details.
3. To meet the needs of the Company, the assignment of staff to the Company shall be authorized, coming from the public administration, legal persons governed by public law or the broader public sector, independent administrative authorities, including legal persons under private law and entities supervised by the government in accordance with Law n° 4440/2016.

Article 22: Privileges Tax Exemptions

1. The company shall enjoy all the administrative, economic and judicial, of both substantive and procedural law, privileges and tax reliefs of a State. Additionally, it shall be exempted from paying duties, direct or indirect taxes, except for Value Added Tax (VAT), the Single Real Estate Tax and the Specific Real Estate Tax, parafiscal charges and any taxes with respect to equipment imported by the Company and intended to fulfill its objectives, in accordance with relevant provisions.
2. Transactions or agreements, concerning any transfer of asset or liability and any real property right to the Company or from the Company to its subsidiaries,

including the registration of such transactions or agreements at the relevant cadastral offices or land registers, are exempted from all taxes other than the Tax on Building Loans and Taxes on the Transfer of Immovable Property, donations, inheritances, where applicable, and of any fee, contribution or rights benefiting the State or any other legal person governed by public law, insurance corporations or third parties. Particularly with regard to the Income Tax, the Company shall be exempted from it, except in case of capital income and gains from capital transfer, in accordance with point α' of Article 46 of Law n° 4172/2013 (A' 167). In any event, Community Law regarding State aids shall apply, with the exception of the issues concerning military applications and matters of national security.

Article 23: Dissolution Liquidation

In the event of the dissolution of the Company, the liquidation shall be carried out in line with the provisions on public limited liability companies. A joint decision of the Minister of Digital Policy, Telecommunications and Information, and any other competent Minister in each case shall regulate specific issues related to the Company's staff, the assets of the Company, movable or immovable, as well as any other matter relating to the dissolution of the Company.

Article 24: Matters of Control, Supervision and Transparency

1. The management and financial control, as well as the supervision of the Company shall be carried out in accordance with its Statutes and Law n° 3429/2005.
2. The Company shall submit within the first trimester of each year, to the Minister of Digital Policy, Telecommunications and Information a report on the objectives achieved and on the expectations for the next year, as well as its revised strategic and operational plan. The first report shall be delivered upon completion of at least one year of the Company's operation.
3. The proposal for the space strategy and the rolling action plan shall be submitted for approval to the Minister of Digital Policy, Telecommunications and Information.
4. The company is obliged to provide all the information related to its functioning or any other relevant information, at the request of the Minister of Digital Policy, Telecommunications and Information.
5. The Administrative Board of the Company may be invited to appear before committees of the Parliament to inform them about any relevant issue in the field of space, without prejudice to matters of national security within the context of military applications and space resources management.

Index

A. Koskina, M. Plionis, *Greek Space Law*, Space and Society,
https://doi.org/10.1007/978-3-031-72813-6

The manufacturer's authorised representative in the EU is Springer Nature Customer Service Centre GmbH, Europaplatz 3, 69115 Heidelberg, Germany. If you have any concerns regarding our products, please contact ProductSafety@springernature.com

Printed and bound by CPI Group (UK) Ltd, Croydon, CR0 4YY

30/09/2026

02225591-0008